Simon Lukas Zeller

Ecology of Genetically Modified Wheat:

AF114010

Simon Lukas Zeller

Ecology of Genetically Modified Wheat:

Performance, Resistance Costs, Mixture Effects and Gene Flow

Südwestdeutscher Verlag für Hochschulschriften

Impressum/Imprint (nur für Deutschland/only for Germany)
Bibliografische Information der Deutschen Nationalbibliothek: Die Deutsche Nationalbibliothek verzeichnet diese Publikation in der Deutschen Nationalbibliografie; detaillierte bibliografische Daten sind im Internet über http://dnb.d-nb.de abrufbar.
Alle in diesem Buch genannten Marken und Produktnamen unterliegen warenzeichen-, marken- oder patentrechtlichem Schutz bzw. sind Warenzeichen oder eingetragene Warenzeichen der jeweiligen Inhaber. Die Wiedergabe von Marken, Produktnamen, Gebrauchsnamen, Handelsnamen, Warenbezeichnungen u.s.w. in diesem Werk berechtigt auch ohne besondere Kennzeichnung nicht zu der Annahme, dass solche Namen im Sinne der Warenzeichen- und Markenschutzgesetzgebung als frei zu betrachten wären und daher von jedermann benutzt werden dürften.
Coverbild: www.ingimage.com

Verlag: Südwestdeutscher Verlag für Hochschulschriften GmbH & Co. KG
Dudweiler Landstr. 99, 66123 Saarbrücken, Deutschland
Telefon +49 681 37 20 271-1, Telefax +49 681 37 20 271-0
Email: info@svh-verlag.de

Approved by: Zürich, Universität Zürich, Dissertation, 2011

Herstellung in Deutschland:
Schaltungsdienst Lange o.H.G., Berlin
Books on Demand GmbH, Norderstedt
Reha GmbH, Saarbrücken
Amazon Distribution GmbH, Leipzig
ISBN: 978-3-8381-2982-2

Imprint (only for USA, GB)
Bibliographic information published by the Deutsche Nationalbibliothek: The Deutsche Nationalbibliothek lists this publication in the Deutsche Nationalbibliografie; detailed bibliographic data are available in the Internet at http://dnb.d-nb.de.
Any brand names and product names mentioned in this book are subject to trademark, brand or patent protection and are trademarks or registered trademarks of their respective holders. The use of brand names, product names, common names, trade names, product descriptions etc. even without a particular marking in this works is in no way to be construed to mean that such names may be regarded as unrestricted in respect of trademark and brand protection legislation and could thus be used by anyone.
Cover image: www.ingimage.com

Publisher: Südwestdeutscher Verlag für Hochschulschriften GmbH & Co. KG
Dudweiler Landstr. 99, 66123 Saarbrücken, Germany
Phone +49 681 37 20 271-1, Fax +49 681 37 20 271-0
Email: info@svh-verlag.de

Printed in the U.S.A.
Printed in the U.K. by (see last page)
ISBN: 978-3-8381-2982-2

Copyright © 2011 by the author and Südwestdeutscher Verlag für Hochschulschriften GmbH & Co. KG and licensors
All rights reserved. Saarbrücken 2011

Ecology of Genetically Modified Wheat:

Performance, Resistance Costs, Mixture Effects and Gene Flow

Simon L. Zeller

Die vorliegende Dissertation wurde von der Mathematisch-naturwissenschaftlichen Fakultät der Universität Zürich im Herbstsemester 2011 als Dissertation angenommen.

Promotionskomitee:
Prof. Dr. Bernhard Schmid (Vorsitz und Leitung)
Prof. Dr. Andrew Hector
Prof. Dr. Beat Keller
Dr. François Felber

I dedicate this thesis to
my parents

CONTENTS

GENERAL INTRODUCTION 7

CHAPTER 1 **Transgene x Environment Interactions in Genetically Modified Wheat**
S.L. Zeller, O. Kalinina, S. Brunner, B. Keller , B. Schmid
(2010) *PLoS ONE* 5(7), e11405 23

CHAPTER 2 **Costs of Resistance in Genetically Modified Wheat**
S.L. Zeller, O. Kalinina, B. Schmid
Manuscript 49

CHAPTER 3 **Mixtures of Genetically Modified Wheat Lines Outperform Monocultures**
S.L. Zeller, O. Kalinina, B. Schmid
Manuscript 77

CHAPTER 4 **Competitive Performance of Pathogen-Resistant Transgenic Wheat: a Phytometer Approach**
O. Kalinina, S.L. Zeller, B. Schmid
Manuscript 103

CHAPTER 5 **Gene Flow of Genetically Modified Wheat**
S. Rieben, S.L. Zeller, O. Kalinina, B. Schmid
Manuscript 133

GENERAL DISCUSSION 167

SUMMARY 189

ZUSAMMENFASSUNG 193

ACKNOWLEDGEMENTS 197

CURRICULUM VITAE 202

GENERAL INTRODUCTION

Fig. 1: Manual harvest of field trial with genetically modified (GM) wheat in 2010

A hungry world

It will be one of mankind's greatest challenges to feed the 9 billion people which are projected to populate our planet by 2050 (Godfray et al. 2010). Already today, it is difficult to meet the global demand for food. Consumption of grains and oilseeds has exceeded production in 7 out of 8 years since 2000 leading to the lowest food stock levels since 1970 (USDA 2008). Spiking food commodity prices made more than 1 billion people in the developing world suffer from hunger in 2008 (FAO 2009). After a brief phase of relieve, food prices in 2011 have again surpassed levels reached in 2008 and were a major cause of popular unrest (FAO 2011).

There are multiple reasons for this worrying development. On the consumption side, the world's population grows by 200'000 hungry mouths everyday, but more importantly, we see sharply increasing demands for protein-rich food, especially in developing countries such as China and India. Furthermore, climate change politics and fossil fuel shortages foster the production of biofuels on lands previously used to grow food (USDA 2008). On the supply side, the annual growth rate in grain and oilseed production has been slowing from 2.2% between 1970 and 1990 to 1.3% since 1990 (USDA 2008). Yields of important staple crops such as wheat have reached a plateau or are even declining in Europe (Brisson et al. 2010) and the United States (Graybosch and Peterson 2010).

Similar to all other internationally traded commodities, prices for agricultural goods increase as soon as demand surpasses supplies. Higher food prices however, will not stop rich countries from importing fodder for their livestock or biofuels. It is mainly the people in the poor countries which are often dependent on food imports that will suffer most. To prevent catastrophic scenarios such as worldwide famines and refugee streams (Cribb 2010), either the demand for food needs to be lowered or the production increased. Furthermore, the distribution and availability of food needs to be improved. As mentioned above, population growth is one of the main drivers of food demand. Attempts to slow down population growth have been made, for example the One-Child policy in China, but it involved drastic violations of human rights. It has to be seen if more social ways can be found to reduce human fertility. The best and certainly healthiest option would be to reduce the world food demand by stop wasting food (Nellemann et al. 2009; Godfray et al. 2010) and limiting meat consumption in the industrialized countries (Vogel 2010). However, there are models that show that even if the industrial countries would cut their meat consumption by half, this would only marginally ease the hunger of the worlds poorest (Stokstad 2010). Ultimately, the world has to try to increase its food production somehow – at least until the population maximum is reached. Demand for food might increase by 70–100% during the next forty years (WorldBank 2007; Royal Society of London 2009). The following paragraph will explain how yields have been improved in the past and discuss factors limiting future efforts.

Need for second green revolution

Doubling the world's food production is very difficult; however it has been done before. The so called "green revolution" allowed increasing the world grain production from 1 to 2 billion tons between 1960 and 2000 (Khush 2001). This major effort was possible due to the introduction of new, improved crop varieties that allowed higher fertilizer and pesticide input and the expansion of cropping area (Evenson and Gollin 2003). Unfortunately, the industrialisation of agriculture led to massive environmental problems (Tilman et al. 2001). One could argue that we need a second "green revolution" but this time, it needs to be sustainable.

Most of the ways used to increase agricultural production in the past were far from being sustainable and will meet severe limitations in the future: firstly, expansion of global agricultural land inevitably means clearing tropical forests and shrubland ecosystems (Gibbs et al. 2010). This leads to increasing greenhouse gas emissions and the loss of biodiversity and important ecosystem services. It is likely, that future environmental policies such as "Reducing Emissions from Deforestation and Degradation" (REDD) will slow down agricultural expansion (Ghazoul et al. 2010). Secondly, more than half of the global food production increase was due to higher fertilizer and pesticide input and therefore dependent on fossil fuels or other non renewable resources (Cordell et al. 2009; Godfray et al. 2010). Since the production maxima for several non-renewable resources is predicted to peak in the near future (Heinberg 2005; Cordell et al. 2009) further intensification of agriculture might not be feasible. Third, 40% of the world's food is currently grown on irrigated fields. However, climate change models predict that many countries are likely to suffer from water scarcity which will negatively affect their agricultural output (Nellemann et al. 2009). Forth, genetic improvements of crop varieties have been a mayor driver of the past green revolution. Such improvements can be achieved by traditional breeding or genetic engineering. The performance and ecology of several novel wheat varieties and lines that contain such genetic improvements will be the topic of this thesis.

The role of transgenic plants

Since the early nineties, ever growing numbers of genetically modified (GM) crops have been released with the promise of superior quality compared to conventional varieties (James 2009). Whereas the adoption of GM crops was fast in countries like USA, Canada, Brazil, Argentina and India, any attempts to release such plants met massive resistance in other places. Besides ethical and religious concerns, opponents of GM technology in Europe criticised that the influence of these plants on human health (Kuiper et al. 2001) and the environment had not been evaluated properly. Key concerns of environmentalist are invasiveness, vertical or horizontal gene flow, other ecological impacts, effects on biodiversity and the impact of presence of GM material in other products (Conner et al. 2003). Some of these concerns have been corroborated by studies showing pollen-mediated gene flow from GM plants into conventional varieties or wild plant species (Quist and Chapela 2001; Mallory-Smith and Zapiola 2008; Zapiola et al. 2008; Piñeyro-Nelson et al. 2009). In Canada, populations of canola plants that have accumulated several herbicide resistance genes were found along highways (Knispel et al. 2008). GM traits obtained trough gene flow from GM fields allow these plants to persist as volunteers in agricultural landscapes (Knispel et al. 2008). Herbicide resistant creeping bentgrass *Agrosis stolonifera* has escaped from field trials and continued to spread for 3 years. Efforts to eliminate these plants proved to be in vain (Zapiola et al. 2008). Furthermore, traces of BT-toxins have been found in streams nearby BT-maize fields (Rosi-Marshall et al. 2007) which may cause unexpected ecosystem-scale consequences. However, no persuasive studies linking GM technology to serious health problems or major environmental disruptions have been published so far. On the opposing side, agrochemical companies promised miracle plants that would solve the food crisis. However, private companies tend to develop profitable crops that reduce the costs of farming rather than enhance yields (USDA 2008). The number of farmers that are willing to grow GM plants containing either BT-toxins against insects, herbicide resistances or both is currently growing fast, especially in developing countries (James 2009). Hence, such plants must be profitable for them (Brookes and Barfoot 2008; James 2009). There is however a heated debate whether the currently available GM crops increase yields or not (Sheridan 2009). Furthermore, strict protection of intellectual property rights has lead to the public perception that the few internationally active agrochemical companies are purely interested in profits and offer no long-term public goods (Godfray et al. 2010). Together, these arguments may have lead to a quasi-moratorium on GM plants in most European countries.

National Research Program 59

In November 2005, 55.7% of 2 million Swiss voters approved a strict five-year moratorium on commercial GM plants (Schläpfer 2008) which was recently prolonged to 2013 without debate. Surprisingly, this vote did not stop biotechnology research in Switzerland. A national research program with the title "Benefits and risks of the deliberate release of genetically modified plants" was installed in spite of the public vote (www.nfp59.ch). The official role of this research program was to provide a scientific basis for a more rational discussion that will contribute to the political decision-making process. Whereas GM plants continue to be unpopular in Switzerland, most people would probably agree that the evaluation of benefits and risks of GM plants by independent researchers is valuable. In 2007 research groups from different Swiss universities, the Federal Institute of Technology in Zurich, the federal agricultural research centres at Reckenholz-Tänikon and Changins-Wädenswil, the Research Institute of Organic Agriculture and private firms started to carry out 29 projects. The aims of these projects were to improve our basic knowledge about biotechnology and environmental interactions, political, social and economic aspects as well as risk assessment issues.

The wheat cluster

Nine of these projects formed an interdisciplinary consortium (www.wheatcluster.ch) that shared GM wheat *Triticum aestivum* L. varieties with fungal resistance genes as model organisms. These were based either on the Mexican spring wheat variety Bobwhite SH 98 26 or the Swiss variety Frisal. The later is a rather old variety that is not cultivated anymore because of low pathogen resistance. Several GM lines were produced by biolistic transformation. Bobwhite GM lines received different alleles of *Pm3* transgenes (Brunner et al. 2011). *Pm3* resistance genes were cloned from hexaploid wheat (Yahiaoui et al. 2004) and confer race-specific resistance to the fungal pathogen powdery mildew *Blumeria graminis* f.sp *tritici*. Frisal GM lines contained either *chitinase* or *chitinase* and *glucanase* transgenes cloned from barley (Bieri et al. 2003). The pathogenesis-related proteins *chitinase* and *glucanase* are known for their broad antifungal effect and its expression should lead to an increased resistance to powdery mildew (Leah et al. 1991; Zhu et al. 1994). More detailed descriptions of the plant material used can be found in the method parts of the Chapters 1–5. Between 2008 and 2011, the research group of the "wheat cluster" carried out field trials at two common field sites which belonged to agricultural research stations in Pully and Zurich Reckenholz.

The nine "wheat cluster" projects aimed to study performance, powdery mildew resistance and interactions with the environment of their model plants (Brunner et al 2011). In addition, biosafety related issues such as gene flow to wild weed species, persistence of volunteer plants and unintended effects on soil bacteria, mycorrhiza, soil macro fauna and arthropods were studied (von Burg et al. 2011; von Burg et al. 2010; Peter et al. 2010). The focus of this thesis, however, lies on the ecological behaviour of GM wheat plants.

Ecological research with GM plants

There are many studies which analysed either the performance or potential risks of GM plants (Wolfenbarger and Phifer 2000; Conner et al. 2003; Rong et al. 2007; Snow et al. 2005) but very few investigated the ecological behaviour of these. One reason for this might be the reluctance of many ecologists to work with GM plants. Depending on the results obtained, even the most independent researchers are likely to endure harsh criticism from either GM-proponents or opponents (Waltz 2009). Furthermore, research with GM plants is cost intensive due to regulatory difficulties, restricted access to plant material and information, and additional expenses for security of field experiments. Nevertheless, research in this field can be rewarding. As we will see in the following chapters, ecological experiments with GM plants can be relevant for concepts of molecular plant breeding as well as for plant evolutionary ecology in general. Plant biotechnology is a relatively new field of research. Hence, the ecological knowledge about GM plants is still scarce and even basic principles have not been proven empirically.

GM plants are supposed to differ only in one or a few intended traits when compared to their genetic background. There have been reports that inserted genes can lead to further, unintended phenotypic changes (Cellini et al. 2004; Snow et al. 2005; Filipecki and Malepszy 2006). However, it is not clear with which frequency such unintended effects appear, if they are relevant for the ecology of the plant and if they are visible all the time or only under certain environmental conditions.

Although it is not the aim of this thesis to perform an ecological risk assessment, (ERA) its results could be relevant for such assessments. Following a tiered approach (Hill and Sendashonga 2003), ERAs of GM plants usually start in climate controlled glasshouses. Only once the safety has been proven, plants can be moved to more realistic field conditions. The question whether potentially harmful unintended effects can be detected already before plants leave the closed glasshouse system should be of central importance for ERAs.

Transformation of GM plants occurs by more or less random integration of transgenes into unpredictable genomic locations through non-homologous recombination (Bock 2001; Stewart et al. 2003). The position of the transgenes in the host plant's genome can however influence its functionality. These so called position effects can reduce or even silence transgenes (Matzke and Matzke 1995). It is not clear to which extent GM lines that are genetically identical, but differ in the position of their transgene, can differ in phenotype and ecology.

Transgene x environment interactions can be interesting for plant evolutionary ecologists that study how single genes affect the ecological and evolutionary behaviour of plants. If wild plant species obtain a new resistance gene, this is usually associated with additional fitness costs (Purrington 2000). In nature, many pathogen resistances are inducible, meaning that they are only activated if a pathogen is present or certain conditions are fulfilled (Heil et al. 2000). Resistance genes in GM plants however, are mostly activated using constitutive promoters leading to continuous gene expression in most plant organs. Great efforts are made to find and use inducible promoters (Corrado and Karali 2009). However, no such plants are currently on the market. Molecular plant breeders generally aim to achieve high transgene expression levels in their plants (Mcbride et al. 1995; Ye et al. 2001). Ecologists would expect that constitutive overexpression of individual genes would cause trade-offs that are detrimental for a plant (Herms and Mattson 1992). In nature, reduced fitness would ultimately lead to the extinction of such plants (Darwin 1859). However, it is not clear if environmental conditions can reduce or even neutralize such fitness costs.

High plant diversity is known to increase biomass of plant communities (Tilman et al. 1996; Hector et al. 1999; Roscher et al. 2005). Ecologists found that complementary and selection effects can explain such patterns. Complementarity effects occur if the performance of mixed species is on average higher than expected from their monoculture yields, while the selection effect explains higher productivity of mixtures by the dominance of individual, highly productive species (Roscher et al. 2005). Complementary effects with plants that differ mainly in their resistance to pathogens have already been described earlier by agronomists (Wolfe 1985; Mundt 2002). They found that if plant species or varieties with differing pathogen resistances are mixed, pathogens cannot spread quickly. Furthermore, diverse plant communities can promote the development of complex pathogen communities with higher stability and lower chance of catastrophic pathogen outbreaks (Haddad et al. 2011). There are examples where mixing different rice varieties resulted in spectacular yield increases (Zhu et al. 2000). However, monoculture is still the main system used in agriculture because it guarantees homogeneous yields and simplifies crop management. Mixtures of phenotypically identical GM lines that differ only in one resistance gene could potentially improve the stability and performance of agricultural systems. We found no published report were this hypothesis was tested.

Ecologists have shown that resource competition, allelopathy and facilitation have important effects on community organization (Callaway and Walker 1997). There is much evidence that competition for resources can have strong effects on survival, growth and reproduction of individual plants (Goldberg 1987). Species with high competitive abilities are more likely to become weeds or invade new habitats (Baker 1974; Sakai et al. 2001). There are fears that GM plants with new traits could potentially become weeds themselves. This seems to have happened with herbicide resistant (HR) canola in Canada and other places (Knispel et al. 2008). HR canola was shown to persist in semi-natural habitats such as road verges (Pessel et al. 2001). However, there are up to date few studies showing escapes and persistence of herbicide resistant crops to natural habitats (Zapiola et al. 2008). This might be because in the case of herbicide resistance, the positive selective value will be restricted to habitats in the agro-ecosystem where the herbicide is applied. This is however not true for other traits. Disease and insect resistances and stress tolerance to cold, drought and salinity could be of use in natural habitats, too (Warwick et al. 2008). It would be of great importance to develop methods that allow assessing the competitive abilities of GM plants with such traits before they are commercialized.

Pollen mediated gene flow among plants and between closely related species is one of the main drivers of evolution and has therefore been studied extensively by evolutionary ecologists (Rieseberg and Burke 2001). One can assume that gene flow of genetically modified plants follows similar principles as in their wild relatives (Ellstrand 2003). Many studies assessed pollen-mediated gene flow from GM field to conventional crops and discussed isolation distances necessary for coexistence of GM and non-GM crops (Rieger et al. 2002; Rong et al. 2007; Mallory-Smith and Zapiola 2008). However, pollen mediated gene flow within the field has not been in the focus of research. In maize, for example, it is known that within one generation transgenes can multiply through gene flow within the field (Dietiker et al. 2011). This can potentially threaten GM threshold limits set by regulators. It is therefore important to study within-field gene flow and how it can be influenced by GM traits or environmental factors.

Thesis Outline

The aim of this thesis is to better understand the ecology of GM wheat. Ecology is commonly defined as the study of relationships between organisms and their environment (Stauffer 1957; Friederichs 1958). We extend this definition by adding the genetic make up of our study organism; in our case the transgenes. This means that we are not only interested in interactions of transgenic plants with their environment but also how plants as organisms cope with additional genes in general. I will start with an assessment of four transgenic wheat lines that were grown with or without additional nutrients and fungicide and grown in the glasshouse or in a field trial (Chapter 1). This set up allows us to study how wheat lines that differ in either the presence or absence of a single transgene or merely the position of this transgene in their genomes, interact with several abiotic environments. Besides traits that are related to performance and pathogen resistance we are also interested in unintended effects that can be caused by the presence of transgenes. We continue with a second experiment that includes four GM lines with resistance against powdery mildew that are grown in presence or absence of this pathogen (Chapter 2). This allows us to explore if additional resistance genes cause additional costs for the individual plant and if such costs depend on the biotic environment. We than start to mix different GM and non-GM wheat lines to see if at the population level, increased diversity can improve pathogen resistance and reduce negative fitness effects caused by the presence of transgenes (Chapters 3). In an additional field trial seedlings of 15 GM and non-GM wheat lines are planted as "phytometers" into plant communities composed of the same 15 lines. The phytometer technique is used for the first time to compare the competitive performance of GM and non-GM wheat lines (Chapter 4). Finally we perform two experiments to assess pollen mediated gene flow of GM and non-GM wheat plants within the field and over short distances (Chapter 5).

In **Chapter 1**, we firstly gain basic knowledge about our model plants by exploring if an additional transgene (*Pm3b*) inserted into the genome of four identical wheat lines leads to the expected increase in resistance to powdery mildew but also if there are unintended effects. Furthermore, it is interesting to know if all transformed lines behave similarly or if there is line-specific variation. We than test if the magnitude or even direction of intended and unintended effects can change with the environment the plants are grown in. Besides fungicide and fertilizer treatments, we grew the plants both in the glasshouse and in a realistic agronomic field. The later allows us to evaluate to which extent data from preceding glasshouse experiments can be used to predict the behaviour of GM plants in the field.

In **Chapter 2**, we investigate if GM plants resistant to powdery mildew suffer from costs of resistance in absence of the pathogen. Individual plants of different GM wheat lines that were either based on the genetic background Bobwhite (*Pm3b* transgene) or Frisal (chitinase or/and glucanase transgene) are grown in plots sprayed with fungicide or naturally or artificially infected with powdery mildew. Furthermore, these plots are made up of plant communities with variable diversity levels to study if higher diversity improves the performance of individual plants.

In **Chapter 3**, three wheat lines, two of them with different *Pm3* alleles effective against different strains of powdery mildew, are grown in monocultures or mixtures of two. In this field trial we hypothesize that higher GM-concentration and higher GM-richness reduces pathogen infection and therefore increases seed yields. Performance measurements are taken on population and individual-level to study underlying ecological mechanisms.

In **Chapter 4**, we use phytometers to study competitive interactions between 15 different GM and non-GM wheat varieties and lines. Besides the main effects of the transgenes we analyse the effects of different nutrient environments (see Chapter 1) and biotic environments (see Chapter 2 and 3) on the performance and pathogen resistance of individual phytometer plants. Furthermore, we assess the interaction of these two factors (transgene x environment). Finally, the differences in behaviour of GM and non-GM lines if planted into their own rather than into different lines as competitive environments are studied. From the applied point of view, this experiment will show if GM plants have the tendency to outcompete other members of the plant community and if phytometers are useful tool to assess the ecology of GM plants in general.

In **Chapter 5**, we study pollen mediated gene flow of GM and non-GM wheat lines. Knowledge about gene flow in GM plants is central for the question of coexistence of GM and non-GM crops but also to understand the influence of individual genes on the reproduction of plants in general. In a first experiment, we replant seeds gained from phytometer plants that grew in plots with differing wheat varieties (Chapter 4) and check if they were cross-pollinated. The data is used to study differences in cross-pollination rates of GM and non GM-varieties. A second experiment measures the decline of cross-pollination rates with increasing distance and if there are differences between GM varieties and environmental influences.

References

Baker, H. G. (1974) The evolution of weeds. *Annual Review of Ecology and Systematics,* **5,** 1–24.

Bieri, S., Potrykus, I. & Futterer, J. (2003) Effects of combined expression of antifungal barley seed proteins in transgenic wheat on powdery mildew infection. *Molecular Breeding,* **11,** 37–48.

Bock, R. (2001) Transgenic plastids in basic research and plant biotechnology. *Journal of Molecular Biology,* **312,** 425–438.

Brisson, N., Gate, P., Gouache, D., Charmet, G., Oury, F.-X. & Huard, F. (2010) Why are wheat yields stagnating in Europe? A comprehensive data analysis for France. *Field Crops Research,* **119,** 201–212.

Brookes, G. & Barfoot, P. (2008) Global impact of biotech crops: socio-economic and environmental effects, 2006. *AgBioForum,* **11,** 21–38.

Brunner, S., Hurni, S., Herren, G., Kalinina, O., von Burg, S., Zeller, S., Schmid, B., Winzeler, M. & Keller, B. (2011) Transgenic Pm3b wheat lines show resistance to powdery mildew in the field. *Plant Biotechnology Journal,* **in press**.

Callaway, R. M. & Walker, L. R. (1997) Competition and facilitation: a synthetic approach to interactions in plant communities. *Ecology,* **78,** 1958–1965.

Cellini, F., Chesson, A., Colquhoun, I., Constable, A., Davies, H. V., Engel, K. H., Gatehouse, A. M., Karenlampi, S., Kok, E. J., Leguay, J. J., Lehesranta, S., Noteborn, H. P., Pedersen, J. & Smith, M. (2004) Unintended effects and their detection in genetically modified crops. *Food and Chemical Toxicology,* **42,** 1089–1125.

Conner, A. J., Glare, T. R. & Nap, J. P. (2003) The release of genetically modified crops into the environment. Part II. Overview of ecological risk assessment. *Plant Journal,* **33,** 19–46.

Cordell, D., Drangert, J. O. & White, S. (2009) The story of phosphorus: global food security and food for thought. *Global Environmental Change-Human and Policy Dimensions,* **19,** 292–305.

Corrado, G. & Karali, M. (2009) Inducible gene expression systems and plant biotechnology. *Biotechnology Advances,* **27,** 733–743.

Cribb, J. (2010) *The coming famine: the global food crisis and what we can do to avoid it.* University of California Press Ltd., London.

Darwin, C. (1859) *On the origin of species by means of natura selection, or the preservation of favoured races in the struggle for life.* Atheneum, New York.

Dietiker, D., Oehen, B., Ochsenbein, C., Westgate, M. E. & Stamp, P. (2011) Field simulation of transgenic seed admixture dispersion in maize with a blue kernel colour marker. *Crop Science,* **51,** doi: 10.2135/cropsci2010.06.0311.

Ellstrand, N. C. (2003) Current knowledge of gene flow in plants: implications for transgene flow. *Philosophical Transactions of the Royal Society B-Biological Sciences,* **358,** 1163–1170.

Evenson, R. E. & Gollin, D. (2003) Assessing the impact of the Green Revolution, 1960 to 2000. *Science,* **300,** 758–762.

FAO (2009) *The state of food and agriculture 2009.* Food and Agriculture Organization of the United Nations, Rome.

FAO (2011) *Global food price monitor.* Food and Agriculture Organization of the United Nations, Rome.

Filipecki, M. & Malepszy, S. (2006) Unintended consequences of plant transformation: a molecular insight. *Journal of Applied Genetics,* **47,** 277–286.

Friederichs, K. (1958) A definition of ecology and some thoughts about basic concepts. *Ecology,* **39,** 154–159.

Ghazoul, J., Butler, R. A., Mateo-Vega, J. & Koh, L. P. (2010) REDD: a reckoning of environment and development implications. *Trends in Ecology & Evolution,* **25,** 396–402.

Gibbs, H. K., Ruesch, A. S., Achard, F., Clayton, M. K., Holmgren, P., Ramankutty, N. & Foley, J. A. (2010) Tropical forests were the primary sources of new agricultural land in the 1980s and 1990s. *Proceedings of the National Academy of Sciences of the United States of America,* **107,** 16732–16737.

Godfray, H. C. J., Beddington, J. R., Crute, I. R., Haddad, L., Lawrence, D., Muir, J. F., Pretty, J., Robinson, S., Thomas, S. M. & Toulmin, C. (2010) Food security: the challenge of feeding 9 billion people. *Science,* **327,** 812–818.

Goldberg, D. E. (1987) Neighborhood competition in an old-field plant community. *Ecology,* **68,** 1211–1223.

Graybosch, R. A. & Peterson, C. J. (2010) Genetic improvement in winter wheat yields in the great plains of North America, 1959–2008. *Crop Science,* **50,** 1882–1890.

Haddad, N. M., Crutsinger, G. M., Gross, K., Haarstad, J. & Tilman, D. (2011) Plant diversity and the stability of foodwebs. *Ecology Letters,* **14,** 42–46.

Hector, A., Schmid, B., Beierkuhnlein, C., Caldeira, M. C., Diemer, M., Dimitrakopoulos, P. G., Finn, J. A., Freitas, H., Giller, P. S., Good, J., Harris, R., Hogberg, P., Huss-Danell, K., Joshi, J., Jumpponen, A., Korner, C., Leadley, P. W., Loreau, M., Minns, A., Mulder, C. P. H., O'Donovan, G., Otway, S. J., Pereira, J. S., Prinz, A., Read, D. J., Scherer-Lorenzen, M., Schulze, E. D., Siamantziouras, A. S. D., Spehn, E. M., Terry, A. C., Troumbis, A. Y., Woodward, F. I., Yachi, S. & Lawton, J. H. (1999) Plant diversity and productivity experiments in European grasslands. *Science,* **286,** 1123–1127.

Heil, M., Hilpert, A., Kaiser, W. & Linsenmair, K. E. (2000) Reduced growth and seed set following chemical induction of pathogen defence: does systemic acquired resistance (SAR) incur allocation costs? *Journal of Ecology,* **88,** 645–654.

Heinberg, R. (2005) *The party's over: oil, war and the fate of industrial society.* New Society Publishers, Gabriola Island, Canda.

Herms, D. A. & Mattson, W. J. (1992) The dilemma of plants – to grow or defend. *Quarterly Review of Biology,* **67,** 283–335.

Hill, R. A. & Sendashonga, C. (2003) General principles for risk assessment of living modified organisms: lessons from chemical risk assessment. *Environmental Biosafety Research,* **2,** 81–88.

James, C. (2009) *Global status of commercialized biotech/GM crops: 2009.* ISAAA, Ithaca, NY.

Khush, G. S. (2001) Green revolution: the way forward. *Nature Reviews Genetics,* **2,** 815–822.

Knispel, A. L., McLachlan, S. M., Van Acker, R. C. & Friesen, L. F. (2008) Gene flow and multiple herbicide resistance in escaped canola populations. *Weed Science,* **56,** 72–80.

Kuiper, H. A., Kleter, G. A., Noteborn, H. P. & Kok, E. J. (2001) Assessment of the food safety issues related to genetically modified foods. *Plant Journal,* **27,** 503–28.

Leah, R., Tommerup, H., Svendsen, I. & Mundy, J. (1991) Biochemical and molecular characterization of three barley seed proteins with antifungal properties. *Journal of Biological Chemistry,* **266,** 1564–73.

Mallory-Smith, C. & Zapiola, M. (2008) Gene flow from glyphosate-resistant crops. *Pest Management Science,* **64,** 428–440.

Matzke, M. A. & Matzke, A. J. M. (1995) How and why do plants inactivate homologous (trans) genes? *Plant physiology,* **107,** 679–685.

Mcbride, K. E., Svab, Z., Schaaf, D. J., Hogan, P. S., Stalker, D. M. & Maliga, P. (1995) Amplification of a chimeric bacillus gene in chloroplasts leads to an extraordinary level of an insecticidal protein in tobacco. *Bio-Technology,* **13,** 362–365.

Mundt, C. C. (2002) Use of multiline cultivars and cultivar mixtures for disease management. *Annual Review of Phytopathology,* **40,** 381–410.

Nellemann, C., MacDevette, M., Manders, T., Eickhout, B., Svihus, B., Prins, A. G. & Kaltenborn, B. P. (2009) *The environmental food crisis – The environment's role in averting future food crises.* United Nations Environment Programme, Cambridge.

Pessel, F. D., Lecomte, J., Emeriau, V., Krouti, M., Messean, A. & Gouyon, P. H. (2001) Persistence of oilseed rape (*Brassica napus* L.) outside of cultivated fields. *Theoretical and Applied Genetics,* **102,** 841–846.

Peter, M., Lindfeld, A. & Nentwig, W. (2010) Does GM wheat affect saprophagous *Diptera* species (*Drosophilidae, Phoridae*)? *Pedobiologia,* **53,** 271–279.

Piñeyro-Nelson, A., Van Heerwaarden, J., Perales, H. R., Serratos-Hernandez, J. A., Rangel, A., Hufford, M. B., Gepts, P., Garay-Arroyo, A., Rivera-Bustamante, R. & Alvarez-Buylla,

E. R. (2009) Transgenes in Mexican maize: molecular evidence and methodological considerations for GMO detection in landrace populations. *Molecular Ecology,* **18,** 750–761.
Purrington, C. B. (2000) Costs of resistance. *Current Opinion in Plant Biology,* **3,** 305–308.
Quist, D. & Chapela, I. H. (2001) Transgenic DNA introgressed into traditional maize landraces in Oaxaca, Mexico. *Nature,* **414,** 541–543.
Rieger, M. A., Lamond, M., Preston, C., Powles, S. B. & Roush, R. T. (2002) Pollen-mediated movement of herbicide resistance between commercial canola fields. *Science,* **296,** 2386–2388.
Rieseberg, L. H. & Burke, J. M. (2001) The biological reality of species: gene flow, selection, and collective evolution. *Taxon,* **50,** 47–67.
Rong, J., Lu, B. R., Song, Z. P., Su, J., Snow, A. A., Zhang, X. S., Sun, S. G., Chen, R. & Wang, F. (2007) Dramatic reduction of crop-to-crop gene flow within a short distance from transgenic rice fields. *New Phytologist,* **173,** 346–353.
Roscher, C., Temperton, V. M., Scherer-Lorenzen, M., Schmitz, M., Schumacher, J., Schmid, B., Buchmann, N., Weisser, W. W. & Schulze, E. D. (2005) Overyielding in experimental grassland communities – irrespective of species pool or spatial scale. *Ecology Letters,* **8,** 576–577.
Rosi-Marshall, E. J., Tank, J. L., Royer, T. V., Whiles, M. R., Evans-White, M., Chambers, C., Griffiths, N. A., Pokelsek, J. & Stephen, M. L. (2007) Toxins in transgenic crop byproducts may affect headwater stream ecosystems. *Proceedings of the National Academy of Sciences of the United States of America,* **104,** 16204–16208.
RoyalSocietyofLondon (2009) *Reaping the benefits: Science and the sustainable intensification of global agriculture.* The Royal Society, London.
Sakai, A. K., Allendorf, F. W., Holt, J. S., Lodge, D. M., Molofsky, J., With, K. A., Baughman, S., Cabin, R. J., Cohen, J. E., Ellstrand, N. C., McCauley, D. E., O'Neil, P., Parker, I. M., Thompson, J. N. & Weller, S. G. (2001) The population biology of invasive species. *Annual Review of Ecology and Systematics,* **32,** 305–332.
Schläpfer, F. (2008) Determinants of voter support for a five-year ban on the cultivation of genetically modified crops in Switzerland. *Journal of Agricultural Economics,* **59,** 421–435.
Sheridan, C. (2009) Report claims no yield advantage for Bt crops. *Nature Biotechnology,* **27,** 588–589.
Snow, A. A., Andow, D. A., Gepts, P., Hallerman, E. M., Power, A., Tiedje, J. M. & Wolfenbarger, L. L. (2005) Genetically engineered organisms and the environment: Current status and recommendations. *Ecological Applications,* **15,** 377–404.
Stauffer, R. C. (1957) Haeckel, Darwin, and ecology. *Quarterly Review of Biology,* **32,** 138–144.
Stewart, C. N., Halfhill, M. D. & Warwick, S. I. (2003) Transgene introgression from genetically modified crops to their wild relatives. *Nature Reviews Genetics,* **4,** 806–817.
Stokstad, E. (2010) Could less meat mean more food? *Science,* **327,** 810–811.
Tilman, D., Fargione, J., Wolff, B., D'Antonio, C., Dobson, A., Howarth, R., Schindler, D., Schlesinger, W. H., Simberloff, D. & Swackhamer, D. (2001) Forecasting agriculturally driven global environmental change. *Science,* **292,** 281–284.
Tilman, D., Wedin, D. & Knops, J. (1996) Productivity and sustainability influenced by biodiversity in grassland ecosystems. *Nature,* **379,** 718–720.
USDA (2008) *Global agricultural supply and demand: Factors contributing to the recent increase in food commodity prices.* United Stated Department of Agriculture, Washington DC.
Vogel, G. (2010) For more protein, filet of cricket. *Science,* **327,** 811–811.
von Burg, S., Müller, C. B. & Romeis, J. (2010) Transgenic disease-resistant wheat does not affect the clonal performance of the aphid *Metopolophium dirhodum Walker. Basic and Applied Ecology,* **11,** 257–263.
von Burg, S., van Veen, F. J. F., Álvarez-Alfageme, F. & Romeis, J. (2011) Aphid-parasitoid community structure on genetically modified wheat. *Biology Letters*, doi: 10.1098/rsbl.2010.1147.
Waltz, E. (2009) Battlefield. *Nature,* **461,** 27–32.

Warwick, S. I., Legere, A., Simard, M. J. & James, T. (2008) Do escaped transgenes persist in nature? The case of an herbicide resistance transgene in a weedy *Brassica rapa* population. *Molecular Ecology,* **17,** 1387–1395.

Wolfe, M. S. (1985) The current status and prospects of multiline cultivars and variety mixtures for disease resistance. *Annual Review of Phytopathology,* **23,** 251–273.

Wolfenbarger, L. L. & Phifer, P. R. (2000) The ecological risks and benefits of genetically engineered plants. *Science,* **290,** 2088–2093.

WorldBank (2007) *World development report 2008: agriculture for development.* The International Bank for Reconstruction and Development / The World Bank, Washington DC.

Yahiaoui, N., Srichumpa, P., Dudler, R. & Keller, B. (2004) Genome analysis at different ploidy levels allows cloning of the powdery mildew resistance gene Pm3b from hexaploid wheat. *Plant Journal,* **37,** 528–538.

Ye, G. N., Hajdukiewicz, P. T. J., Broyles, D., Rodriguez, D., Xu, C. W., Nehra, N. & Staub, J. M. (2001) Plastid-expressed 5-enolpyruvylshikimate-3-phosphate synthase genes provide high level glyphosate tolerance in tobacco. *Plant Journal,* **25,** 261–270.

Zapiola, M. L., Campbell, C. K., Butler, M. D. & Mallory-Smith, C. A. (2008) Escape and establishment of transgenic glyphosate-resistant creeping bentgrass *Agrostis stolonifera* in Oregon, USA: a 4-year study. *Journal of Applied Ecology,* **45,** 486–494.

Zhu, Q., Maher, E. A., Masoud, S., Dixon, R. A. & Lamb, C. J. (1994) Enhanced protection against fungal attack by constitutive coexpression of chitinase and glucanase Genes in transgenic tobacco. *Bio-Technology,* **12,** 807–812.

Zhu, Y., Chen, H., Fan, J., Wang, Y., Li, Y., Chen, J., Fan, J., Yang, S., Hu, L., Leung, H., Mew, T. W., Teng, P. S., Wang, Z. & Mundt, C. C. (2000) Genetic diversity and disease control in rice. *Nature,* **406,** 718–722.

Figure 1: Photograph taken by Simon Zeller

CHAPTER 1

Transgene x environment interactions in genetically modified wheat

S.L. Zeller, O. Kalinina, S. Brunner, B. Keller, B. Schmid (2010), *PloS ONE*, **5**, e11405

Fig. 2: Spike of a GM Bobwhite line (*Pm3b*#2) infected with ergot *Claviceps purpurea*

Abstract

The introduction of transgenes into plants may cause unintended phenotypic effects which could have an impact on the plant itself and the environment. Little is published in the scientific literature about the interrelation of environmental factors and possible unintended effects in genetically modified (GM) plants.

We studied transgenic bread wheat *Triticum aestivum* lines expressing the wheat *Pm3b* gene against the fungus powdery mildew *Blumeria graminis* f.sp. *tritici*. Four independent offspring pairs, each consisting of a GM line and its corresponding non-GM control line, were grown under different soil nutrient conditions and with and without fungicide treatment in the glasshouse. Furthermore, we performed a field experiment with a similar design to validate our glasshouse results.

The transgene increased the resistance to powdery mildew in all environments. However, GM plants reacted sensitive to fungicide spraying in the glasshouse. Without fungicide treatment, in the glasshouse GM lines had increased vegetative biomass and seed number and a twofold yield compared with control lines. In the field these results were reversed. Fertilization generally increased GM / control differences in the glasshouse but not in the field.

Two of four GM lines showed up to 56% yield reduction and a 40-fold increase of infection with ergot disease *Claviceps purpurea* compared with their control lines in the field experiment; one GM line was very similar to its control.

Our results demonstrate that, depending on the insertion event, a particular transgene can have large effects on the entire phenotype of a plant and that these effects can sometimes be reversed when plants are moved from the glasshouse to the field. However, it remains unclear which mechanisms underlie these effects and how they may affect concepts in molecular plant breeding and plant evolutionary ecology.

Introduction

The widespread use of genetically modified (GM) plants in agriculture, together with the growing number of different crop species and introduced genes, demands sound environmental risk assessment (Wolfenbarger and Phifer 2000; Conner et al. 2003; Cellini et al. 2004; Snow et al. 2005). Following a tiered approach (Hill and Sendashonga 2003), data from such preliminary risk assessment usually form the basis for extended field trials or lead to the rejection of GM plants from further testing at an early stage (Conner and Christey 1994). Such studies often focus on the risk that a transgene may not show the desired phenotypic effect if the GM plants are moved from the controlled glasshouse environment to the more variable field conditions. However, few studies have reported potentially unintended phenotypic effects of transgenes in GM plants exposed to a range of realistic environmental conditions (Purrington and Bergelson 1995; Gertz et al. 1999). From evolutionary and ecological studies on wild plants it is well known that genotype × environment interactions can be large (Schlichting 1986; Sultan 1987; Schmid 1992; Sultan 2001; Yahiaoui et al. 2004), suggesting that similar interactions might occur in GM plants exposed to different environments, including glasshouse versus field environments. Plant breeders know intuitively that plant performance needs to be tested in realistic agricultural environments and regulatory authorities demand such assessments in their guidelines (EFSA 2006). Recent studies compared metabolic composition and transcriptional changes in GM Maize grown among environments and *in vitro* and outdoors (Coll et al. 2009; Barros et al. 2010). They found that differences between GM and control plants in metabolic profiles observed under standardized laboratory conditions were lost in the field. However, whether the same was true for ecological traits was not reported in these studies. Furthermore, a careful search in the literature for replicated and randomized studies about the ecological behaviour of GM and control plants in glasshouse versus field environments did not return any published references.

We therefore used the spring wheat variety Bobwhite SH 98 26 *Triticum aestivum* L. — transformed with the wheat *Pm3b* powdery mildew resistance gene (Yahiaoui et al. 2004) — as a model system to study potential transgene × environment interactions in genetically modified plants. We grew four offspring pairs, each consisting of a GM line and its corresponding non-GM control line under different soil nutrient conditions and fungicide treatment in the glasshouse and the field. Although well studied and not showing any abnormalities in the glasshouse, these plants had never been planted outdoors prior to our experiments. We investigated to what extent the single inserted transgene could influence the disease resistance and overall fitness of our study plants and how these effects were modified by moving the plants from the glasshouse to the field. Since the germination rate of our plants was close to 100% (S. Zeller, unpublished data), agronomical performance traits such as seed yield and seed number were used to indirectly assess changes in plant fitness (Haldane 1927). We asked the following questions: (i) Does the transgene enhance resistance to powdery mildew *B. graminis* f.sp. *tritici* (DC.) Speer and does it have other phenotypic effects such as fitness costs? (ii) Do we find these effects in all transformed lines or is there line-specific variation? (iii) Can intended and unintended effects of the transgene be influenced by environmental factors and are such effects detectable both in the glasshouse and in the field? We consider this study both as an example of how the ecological behaviour of genetically modified plants can be studied with experimental approaches and how such research can lead to insights into phenotypic effects of inserting a single gene artificially into a plant.

Materials and Methods

Genetically Modified Wheat

We used four wheat lines carrying the transgene *Pm3b* in different position on the genome and their respective non-transgenic control lines (null-segregants), each derived from different transformation events (von Burg et al. 2010; Peter et al. 2010). *Pm3b* confers race-specific resistance to powdery mildew and was cloned from hexaploid wheat (Yahiaoui et al. 2004). The lines were generated by biolistic transformation of spring wheat variety Bobwhite SH 98 26 (Pellegrineschi et al. 2002). The plasmids pAHC17+NotI (*PMI*) and pAHC17+3NotI (*Pm3b*) were used as vectors (Christensen and Quail 1996; Travella et al. 2006). After *Not*I (for *Pm3b*) or *Not*I/*Hin*dIII (for *PMI*) digestion, only the desired fragments, but no vector sequences, were co-bombarded into wheat. The *Pm3b* gene was cloned under the control of the *Zea mays* L. (maize) ubiquitin promoter (Christensen and Quail 1996) and transformants were selected on mannose-containing media using the phosphomannose isomerase (PMI)-coding gene as selectable marker (Reed et al. 2001). After regeneration of T0 transformants, four independent T1 families were selected. From each T1 family, an offspring pair was further propagated consisting of a homozygous transgenic plant (GM lines *Pm3b*#1–4) and a null-segregant, i.e. a plant that did neither inherit the *Pm3b* transgene nor the selectable marker (control lines S3b#1–4). Absence/presence of the transgenes was confirmed by Southern hybridization analysis (Southern 2006) using probes from the PM3B (bp 1231–1956 as referred to the GenBank accession AY325736) and PMI (bp 271–810 as referred to the GenBank accession AAC74685) encoding region. The GM lines contained the *Pmi* gene as well as one complete copy of *Pm3b*, and in the case of Pm3b#4 an additional fragment, which segregated as a single Mendelian locus in the T1 generation. The null-segregants did not show any hybridization signal with the probes from the *Pm3b* as well as the *Pmi* coding genes. For both transgenic as well as null-segregant lines we can not exclude the presence of fragments from the coding genes or promoter/terminator regions which were not covered by the probes used in Southern blotting. The offspring pairs were multiplied to T4 and used for the glasshouse and field experiments. The seeds used in this study were thus obtained from GM and control lines that had passed through four generations of sexual reproduction. Studies with *Drosophila melanogaster* (Henikoff 1979) and *Saccharomyces cerevisiae* (Gottschling et al. 1990) showed that a gene's position on the chromosome can influence its expression. We therefore assessed the expression level of the *Pm3b* transgene in the four GM lines by semi-quantitative RT-PCR using RNA isolated from leaves of seedlings grown in the glasshouse (Figure S1). As control for equal amount and quality of template cDNA, the expression levels of the *Mlo* gene (Yu et al. 2005) were determined.

Glasshouse Experiment

The glasshouse experiment took place in a climate-controlled glasshouse at the Institute of Evolutionary Biology and Environmental Studies, University of Zurich, Switzerland, from August 2007 to February 2008 (day/night temperature: 21/16 C°; additional light: 14 h/10 h day/night period, daily watering by hand). Seedlings of each line were planted individually into 11 cm square pots containing sterilized soil (Ökohum lawn soil, Ökohum AG, Herrenhof, Switzerland). The design consisted of the four GM and the four control wheat lines crossed with three soil nutrient levels (0, 1 or 2 g of "Osmocote exact mini" per L; Scotts, Waardenburg, The Netherlands). One gram of Osmocote per L corresponded to 13.2 g N, 6.6 g P, 9.1 g K and 1.7 g Mg m^{-2}. Natural infection of the wheat plants by powdery mildew occurred 1 month after planting. One half of the experiment was subsequently sprayed with a systemic fungicide specific to mildew (2 ml l^{-1} Opus Top; 83.7 g l^{-1} Epoxiconazol and 250 g l^{-1} Fenpropionazol; Maag Agro AG, Dielsdorf, Switzerland). The active ingredient epoxiconazol blocks fungal cell pathways and activates the plants pathogen defences whereas fenpropionazol blocks two enzymes that are related to the fungal cell-wall synthesis. We used a high fungicide concentration (2ml/l); this caused slight leaf chlorosis on several plants that disappeared after a few days. All tested lines were affected equally. Each of the 8 x 3 line-by-nutrient level combinations was replicated five times. Plants were harvested 162 days after the start of the experiment.

Field Experiment

The field experiment took place at an agricultural research station in Zurich-Reckenholz, Switzerland. It started in March 2008 and lasted until August 2008. Four replicate blocks, each with sixteen 1 x 1.08 m plots, were sown with seeds of the same eight wheat lines as used in the glasshouse experiment. In each plot, 400 seeds were sown in six rows with a distance of 18 cm between rows using an Oyjord plot drill system (Wintersteiger AG, Ried, Austria). Fertilizer was applied at the phenological stage 11 and 39 (Zadoks et al. 1974) to half of the plots (two times 3 g N m^{-2} as "Ammonsalpeter 27.5", Lonza, Visp, Switzerland).

The natural field soil provided the plants with sufficient phosphorous, potassium and magnesium (80, 235 and 234 mg kg^{-1}). All plots were sprayed with the herbicide cocktail Concert SX (40% Thifensulfurone, 4% Metusulfurone-methyl; Stähler Suisse AG, Zofingen, Switzerland) and Starane super (120 g l^{-1} Bromoxynil, 120 g l^{-1} Ioxynil, 100 g l^{-1} Fluroxypyr-metilheptil-ester; Omya Agro AG, Safenwil, Switzerland) in the beginning of May. In each plot, five individual plants were marked shortly after germination. Powdery mildew and ergot *Claviceps purpurea* (FR.) TUL. infection occurred naturally. Vandals damaged 53 of the 64 plots at random by removing the tops of some plants early in the flowering stage. The damage-induced loss of leaf area was within the natural variation observed in the field and smaller than the herbivory caused by *Oulema melanopus* L. (cereal leaf beetle). The damaged plots recovered within 2–3 weeks and regained their original height and vegetative mass. We recorded the exact area of damage within each plot and replaced all marked plants that had suffered damage (46.3%). A second field experiment with the same plant lines was carried out in an adjacent field the following year. Although plants grew higher because of more favourable weather conditions, the different wheat lines performed very similar as in the 2008 trials (S. Zeller *et al.*, unpublished data). We are therefore confident that the here presented results and conclusions were not influenced by this disturbance.

Response Variables

We assessed the degree of powdery mildew infection (Eyal et al. 1987) and the phenological stage (Zadoks et al. 1974) 80 days after planting. Plants with visible powdery mildew colonies on all their leaves (including flag leaf) were considered infected. We defined plant height as the highest point of the plant measured from the soil and recorded it at the end of the growing season. For these three variables, powdery mildew infection, phenological stage and plant height, we used the maximum values of all tillers per pot or of the five marked plants per plot in glasshouse or field experiment, respectively, for analysis. After ripening, all plants were cut at ground level and separated into vegetative and reproductive parts (spikes). These were then dried at 80 and 25 C°, respectively, and weighed. We then threshed the reproductive parts, counted and removed the seeds infected by ergot (only in field trial) and obtained the total seed mass which is equivalent to the seed yield. The seed number was calculated from the seed yield divided by the average seed mass. The latter was determined on a sample of seeds, one spike in the glasshouse or 1,000 seeds from all spikes in each 1 x 1.08 m plot in the field. The vegetative mass, seed number and seed yield were total measurements of all plants growing in a pot or a plot. Ergot infection rate was calculated as percentage of seed number.

Data Analysis

In a factorial design, we grew the eight wheat lines under different fertilizer treatments (three levels in the glasshouse and two in the field). There were five blocks in the glasshouse and four in the field. We analysed the data of both experiments separately and in combination by analysis of variance (ANOVA). The critical significance level was 0.05 in all analyses. All quantitative pot data from the glasshouse were multiplied by 82.64 to equal an area of 1 m^2. Quantitative field data were divided by 1.08 for the same reason. Regression analysis showed that two variables were slightly affected by the act of vandalism (seed yield: $R^2 = 0.167$ and seed number: $R^2 = 0.094$; $n = 64$). We removed this effect by multiplying the data of the damaged plots with the negative slope from the regression analysis multiplied by the degree of damage (for 10% damaged area: seed yield: –1.003 g; seed number: –20.8). We used the statistical software GenStat (VSN International Ldt.) to fit multiple regression models and summarize the results in ANOVA tables for all variables except powdery mildew infection (see Tables S1–S3). Residual plots were examined to identify outliers and to check if the assumptions of normality and homoscedasticity were fulfilled. The vegetative mass of one unusually heavy plant was identified as an outlier and excluded from the analysis. Phenological stage was transformed to the fourth power (y^4); vegetative mass, seed yield and seed number were square-root transformed; and ergot infection rate was cube-root transformed. The binary mildew infection data were analysed using multiple logistic regression with analysis of deviance (McCullagh and Nelder 1989).

Results

Glasshouse Experiment

One half of the replicates in the glasshouse experiment were sprayed with fungicide to simulate environments with and without powdery mildew. While the control lines benefited from the fungicide treatment, the GM lines reacted negatively ($P<0.001$ for GM/control x fungicide interaction). The yield of the GM lines dropped lower than the yield of the sprayed control lines (Figure 1). This indicates that the cost of resistance might be high if the pathogen is absent. Furthermore, sprayed plants showed an acute stress reaction in form of chlorotic leaves. We decided therefore to exclude the sprayed portion of the experiment from further analysis.

The *Pm3b* transgene had the desired phenotypic effect and increased resistance to powdery mildew in the glasshouse experiment (Figure 1; $P<0.001$ for difference GM/control plants, see Table S1). The yield of the GM lines doubled (from 1.60 to 3.23 tonnes per ha^{-1}) compared to the susceptible control lines. GM plants had also more seeds and higher vegetative biomass than control plants in the glasshouse (Figure 2; both $P<0.001$; see Table S2). Phenological development and plant height were not affected by the transgene, indicating that these traits may be genetically more constrained than the other traits.

The four offspring pairs differed significantly from one another in the five fitness-related traits (phenological stage: $P<0.001$, plant height: $P<0.001$, vegetative mass: $P=0.006$, seed number: $P=0.004$, seed yield: $P=0.014$ for main effect of offspring pair). Alternatively, we tested if there was a significant difference between the four control lines. They differed indeed in all traits except the mildew resistance (phenological stage: $P<0.001$, plant height: $P<0.001$, vegetative mass: $P<0.001$, seed number: $P<0.001$, seed yield: $P<0.001$ for the contrast among offspring lines within control). These differences may be caused by the callus culturing of GM and control lines or effects of the transformation itself. Heritable effects acquired in cell culture can have a genetic basis and plants with such effects are sometimes used in plant breeding (Larkin and Scowcroft 1981; Jones 2005).

Depending on the offspring pair, the inserted transgene had significantly different effects on three of the measured traits (Figure 2B; vegetative mass: P=0.012, seed number: P<0.001, seed yield: P<0.001 for GM/control × offspring pair interaction). This suggests that unintended phenotypic effects of the transgene depended on the location where it had been inserted into the genome. In absolute numbers, line *Pm3b*#4 had the highest yield (4.19 tonnes per ha^{-1}) of the four tested GM lines and proved to be highly resistant to powdery mildew (only 20% of plants infected).

Fertilizer application in the glasshouse had positive effects on all traits except phenological stage (Figure 2A). Fertilization also increased mildew infection (P=0.016) which might be due to the increased growth rate of the host plant (Last 1953). Increased nutrient content of the plant material could have boosted the spread of mildew directly (Bainbridge 1974). Differences between GM and control plants generally increased with nutrient level (vegetative mass: P=0.035, seed number: P<0.001, seed yield: P<0.001 for fertilizer × GM/control interaction). We currently have no explanation for this result which demonstrates the importance of testing effects of transgenes across a range of environments.

Field Experiment

We measured the same traits in the field experiment as in the glasshouse experiment. In addition we recorded infection by ergot fungus, which occurred naturally in the field but not in the glasshouse. Again, we compared first the four GM lines (Pm3b#1–4) with the control lines (S3b#1–4), then the offspring pairs among each other and finally tested the interaction between these two main effects. GM plants with the *Pm3b* transgene showed increased resistance to powdery mildew (Figure 3A and B; P<0.001; see Table S1). In contrast to the glasshouse findings, GM plants had significantly fewer seeds and lower seed yield than control plants (Figure 3A; both P<0.001; see Table S3). Phenological stage, plant height and vegetative mass were not affected by the transgene. In the field, GM plants showed increased infection by ergot fungus compared with control plants (Figure 4; P<0.001).

The four offspring pairs differed in seed number and their level of ergot infection (seed number: P=0.004, ergot infection: P<0.001 for main effect of offspring pair). Effects of the inserted transgene differed among the four offspring pairs for the dependent variables powdery mildew resistance, ergot infection, seed number and seed yield as reflected in significant GM/control × offspring pair interactions (Figure 3B; powdery mildew infection: $P=0.022$; ergot infection: $P<0.001$; seed number: $P<0.001$, seed yield: $P<0.001$). That is, in the field, yields of the GM lines *Pm3b*#2 and #4 were reduced by 56% and 48%, respectively, when compared with the corresponding control lines within offspring pairs. The lines *Pm3b*#2 and #4 were completely resistant to powdery mildew in the field, whereas 12.5% of the *Pm3b*#1 and #3 plants were infected. The difference in ergot infection between GM and control lines was small in offspring pair 1 (Figure 4), moderate in offspring pair 3, and large in offspring pairs 2 and 4. Seed infection rates of around 1 %, as found in lines 2 and 4, can reduce grain quality.

In the field, fertilization increased plant height ($P=0.006$), vegetative mass ($P=0.003$), seed number ($P<0.001$) and seed yield ($P<0.001$). The development of the plants (phenological stage) was not affected by fertilizer application. Similar to the glasshouse, mildew infection increased with fertilizer application in the field ($P<0.001$). However, in contrast to the glasshouse, fertilization did not alter the difference between the GM and control lines in the field.

Comparison between Glasshouse and Field Experiment

To test if the observed differences in transgene effects between glasshouse and field were statistically significant we also analyzed the datasets from the two experiments together, considering the medium and high nutrient levels in the glasshouse as equivalent to the low and high levels in the field, respectively. As expected, glasshouse and field environments differed significantly from each other. Powdery mildew seemed to favour glasshouse conditions which lead to a stronger infection of the plants in the glasshouse than in the field ($P<0.001$) thus increasing the potential benefits of resistance caused by the transgene in the glasshouse. Glasshouse plants developed more slowly (phenological stage: $P<0.001$) and invested slightly more into vegetative mass ($P=0.042$) but had fewer seeds ($P<0.001$) and lower seed yields ($P<0.001$) than field plants.

GM plants had a fitness advantage over control plants in the glasshouse, but a disadvantage in the field (vegetative mass, seed number and seed yield: $P<0.001$, plant height: $P<0.05$ for glasshouse/field × GM/control interaction). While the differences between glasshouse and field could not be assigned to a single environmental factor, the different fertilizer treatments (three levels in the glasshouse and two in the field) did represent such a controlled environmental gradient. We found that fertilizer had similar phenotypic effects in glasshouse and field environments.

Discussion

Transgene × Environment Interactions

This study demonstrates that GM plants can differ in morphological, fitness- and pathogen-related traits from their control plants. We found several significant transgene (GM vs. control) × environment interactions; that is, depending on the environmental conditions the studied transgene against mildew infection had beneficial or detrimental effects on most of the investigated plant traits. GM plants generally benefited from glasshouse conditions with high mildew infection pressure when compared with control plants but showed a stress reaction when powdery mildew was absent due to fungicide spraying. It is possible that the GM plants lacked the energy to cope with the stress caused by this treatment or the chemical itself could have interacted with the transgene or with pathways involved in *Pm3b*-mediated resistance. It is conceivable that the high fungicide dose increased the extent of the stress reaction of GM plants.

Similar to the fungicide treatment in the glasshouse, the natural conditions outdoors seemed to have stressed the GM plants in the field to the extent that their fitness was significantly reduced. Possible causes of environmental stress in the field were drought and neighbour competition. The only deliberately manipulated factor, i.e. fertilizer application, modified the transgene effects only in the glasshouse but not in the field. Apparently the transgene only offered a relative fitness benefit to GM plants growing under conditions of high mildew incidence but low levels of other stresses. These were exactly the conditions met in the glasshouse but not in the field (nor in the glasshouse after fungicide application). Under less beneficial conditions, the GM plants may have paid a physiological cost for the high intrinsic mildew resistance (Bergelson and Purrington 1996).

Differences among GM Lines

The four GM lines, which each contained a single copy of the identical transgene in homozygous condition, differed significantly from each other. There are several potential reasons for these differences. It is possible that cell culturing caused somaclonal variation among the four offspring pairs which subsequently might have interacted differentially with the transgene (Jones 2005). Although theoretically possible (Cubas et al. 1999) we would not expect that such interactions would be stably inherited over five plant generations as we found it here. It seems unlikely that random somaclonal events would cause similar effects in two of the four independently transformed lines (*Pm3b*#2 and #4). A more plausible explanation for the differential effects of the inserted transgene among the four offspring pairs may be that positional effects caused the line-specific differences. Several processes are known to cause such effects (Filipecki and Malepszy 2006). Firstly, an inserted transgene may disrupt native genes. Because spring wheat is hexaploid, consists of more than 80% repetitive, non-genic DNA sequences and each GM line was created by a single insertion event, it is unlikely that the disruption of coding genes or their regulatory sequences could have caused these differential effects (Slade et al. 2005; Dubcovsky and Dvorak 2007). Secondly, the insertion position of a transgene into the genome may have affected its expression level. Studies have shown that transgene expression rates and activity patterns of independently transformed wheat lines with constitutive ubiquitin promoters can vary(Stoger et al. 1999). Depending on the insertion site, flanking DNA regions may partially silence the inserted promoter. Head-to-tail arrangements of the transgenes, in our case of the *Pm3b* and the selectable marker gene, could also have a negative influence on the promoter activity (Rooke et al. 2000). It is also possible that in some lines the transgene was inserted into a region of the genome with low transcription activity (Stam et al. 1998).

The semi-quantitative expression analysis (Figure S1) indicated that the expression of the *Pm3b* transgene did differ between the four GM lines. Thus, although we lack confirmation by quantitative expression data, it appears that the two GM lines *Pm3b*#2 and #4, where the transgene showed the strongest phenotypic effects, also had the strongest transgene expression. Obviously, this hypothesis should be tested with a much larger number of lines differing in expression levels. However, such a study currently would be beyond our capacities to obtain funding and permissions for field trials. If the hypothesis could be confirmed, there would still be the question whether the overexpression of the transgene led to an overabundance of its protein product and the subsequent phenotypic effects or if other mechanisms would be involved.

Besides the quantitative reduction of fitness, we observed that some spikes of the two lines *Pm3b*#2 and #4 also differed in their morphology during flowering time and that the same two lines were also more heavily infected by ergot fungus than the other two GM lines and the four control lines. The altered spike morphology may have increased the likelihood of ergot spores entering the florets (Waines and Hegde 2003). However, no indications of altered spike morphology were observed in the glasshouse.

Implications for Molecular Plant Breeding

Although transgenic plant lines with unintended phenotypes commonly arise during molecular plant breeding (Snow et al. 2005; Filipecki and Malepszy 2006) they can usually be detected earlier and more easily and are thus not further investigated (Cellini et al. 2004) and published. The development of commercial GM plants is based on long selection processes that start in the glasshouse and end in the field. Enormous numbers of seedlings are already discarded before they are exposed to realistic field settings. Our results may have implications for molecular plant breeding: some of the best GM lines in the glasshouse may still show aberrant performance in the field and some not so promising GM lines in the glasshouse may actually be the best for the field. They would likely be lost at early stages of a selection process only targeted at maximum performance under a particular environment. Based on our glasshouse findings, line *Pm3b*#1would have suffered this fate yet was the best in the field. One lesson from our study and from genotype × environment studies in general (Schlichting 1986; Sultan 1987; Schmid 1992; Joshi et al. 2001) is that lines which perform particularly well in a specific environment may pay a cost of specialization and perform poorly in other environments.

Conclusions

Our study demonstrates that inserting a single transgene into the hexaploid wheat genome, along with the desired target effect such as mildew resistance in the present case, can significantly affect other phenotypic traits and thus, as in our case, change the ecological behaviour of the species (hypothesis (i) in Introduction). Such unintended effects of single genes to our knowledge are always smaller in experiments using naturally occurring genetic variation and wild plants (Kingsolver et al. 2001; Tian et al. 2003). Even when we included crop plants, we could not find any publications where single genes reduced quantitative fitness traits in a plant as strongly as in the present case, yet only in the field and not in the glasshouse (Brown 2002). Commercial glyphosate-resistant soybean cultivars were found to suffer from a 5% yield depression that might be caused by the transgene or its insertion process (Elmore et al. 2001). One study tested wheat varieties with introduced resistance genes against leaf and stripe rust and reported a 12% reduction of yield (Griffey and Allan 1986), which was considered to be a very large effect (Ortelli et al. 1996). Compared with these, the yield reductions of 48 and 56% observed in our two GM lines of wheat expressing the *Pm3b* transgene are much larger (Figure 3B).

We found that the level of mildew resistance as well as the magnitude of other phenotypic effects varied significantly between different GM lines (hypothesis (ii) in Introduction). We hypothesize that this variation in phenotypic effects may be due to different expression levels of the *Pm3b* transgene which in turn might have been caused by different insertion positions of the transgene in the genome. Some plant breeders suggest not selecting for plant lines with complete pathogen resistance because costs of such a resistance often outweigh benefits (Brown 2002). In our case this would speak for selecting GM lines with relatively low expression levels yet still increased mildew resistance, i.e. line *Pm3b*#1 (Masci et al. 2003). However, to test the hypothetical correlation between expression level and phenotypic effects would require specific experiments with a larger number of GM lines as used here. With regard to risk assessment our findings are in agreement with the view that each GM line should be tested in a case-by-case approach (Andow and Zwahlen 2006).

Finally, our results show that even if desired phenotypic effects of a transgene are found across a range of environments in a glasshouse experiment, some of these effects can be reversed if GM lines are exposed to natural environmental variation in the field (hypothesis (iii) in Introduction). Although it is likely that commercial plant breeders know of the presence of transgene × environment interactions, it seems that such observations so far have not found their way into the scientific literature. Breeding trials to select lines for further investigation do not need full replication and randomization, yet for an assessment of the ecological behaviour of such lines, replicated and randomized ecological experiments would be required. Our study may serve as an example of potential results that can be obtained in such experiments. We believe that such experiments can help us to gain a deeper understanding of single-gene effects in plant ecology and evolution.

Acknowledgments

We thank F. Meins, A. Hector, N. Waser, J. Weiner, Y. Willi, I. Baldwin, P. Barnett, J. Petermann, D. Gregorowius and Y. Hautier for discussions and comments, the national research station Agroscope Reckenholz-Tänikon ART for setting up the field experiment, M. Nuñez-Marce and I. Kostetskyi for volunteering and numerous helpers in the field for assistance. The helpful comments of three anonymous reviewers were greatly appreciated. This project was supported by the Swiss National Science Foundation and is a part of the wheat-cluster.ch, a sub-unit of the national research programme NRP 59 (SNF 405940-115607).

References

Andow, D. A. & Zwahlen, C. (2006) Assessing environmental risks of transgenic plants. *Ecology Letters,* **9,** 196–214.

Bainbridge, A. (1974) Effect of nitrogen nutrition of host on barley powdery mildew. *Plant Pathology,* **23,** 160–161.

Barros, E., Lezar, S., Anttonen, M. J., van Dijk, J. P., Rohlig, R. M., Kok, E. J. & Engel, K. H. (2010) Comparison of two GM maize varieties with a near-isogenic non-GM variety using transcriptomics, proteomics and metabolomics. *Plant Biotechnology Journal,* **8,** 436–451.

Bergelson, J. & Purrington, C. B. (1996) Surveying patterns in the cost of resistance in plants. *American Naturalist,* **148,** 536–558.

Brown, J. K. M. (2002) Yield penalties of disease resistance in crops. *Current Opinion in Plant Biology,* **5,** 339–344.

Cellini, F., Chesson, A., Colquhoun, I., Constable, A., Davies, H. V., Engel, K. H., Gatehouse, A. M., Karenlampi, S., Kok, E. J., Leguay, J. J., Lehesranta, S., Noteborn, H. P., Pedersen, J. & Smith, M. (2004) Unintended effects and their detection in genetically modified crops. *Food and Chemical Toxicology,* **42,** 1089–125.

Christensen, A. H. & Quail, P. H. (1996) Ubiquitin promoter-based vectors for high-level expression of selectable and/or screenable marker genes in monocotyledonous plants. *Transgenic Research,* **5,** 213–218.

Coll, A., Nadal, A., Collado, R., Capellades, G., Messeguer, J., Mele, E., Palaudelmas, M. & Pla, M. (2009) Gene expression profiles of MON810 and comparable non-GM maize varieties cultured in the field are more similar than are those of conventional lines. *Transgenic Research,* **18,** 801–808.

Conner, A. J. & Christey, M. C. (1994) Plant-breeding and seed marketing options for the introduction of transgenic insect-resistant crops. *Biocontrol Science and Technology,* **4,** 463–473.

Conner, A. J., Glare, T. R. & Nap, J. P. (2003) The release of genetically modified crops into the environment. Part II. Overview of ecological risk assessment. *Plant Journal,* **33,** 19–46.

Cubas, P., Vincent, C. & Coen, E. (1999) An epigenetic mutation responsible for natural variation in floral symmetry. *Nature,* **401,** 157–161.

Dubcovsky, J. & Dvorak, J. (2007) Genome plasticity a key factor in the success of polyploid wheat under domestication. *Science,* **316,** 1862–1866.

EFSA, E. F. S. A. (2006) Guidance document of the scientific panel on genetically modified organisms for the risk assessment of genetically modified plants and derived food and feed. *EFSA Journal,* doi:10.2903/j.efsa.2006.99.

Elmore, R. W., Roeth, F. W., Nelson, L. A., Shapiro, C. A., Klein, R. N., Knezevic, S. Z. & Martin, A. (2001) Glyphosate-resistant soybean cultivar yields compared with sister lines. *Agronomy Journal,* **93,** 408–412.

Eyal, Z., Scharen, A., Prescott, J. & van Ginkel, M. (1987) *The septoria diseases of wheat: Concepts and methods of disease management.* International Maize and Wheat Improvement Center, D.F. Mexico.

Filipecki, M. & Malepszy, S. (2006) Unintended consequences of plant transformation: a molecular insight. *Journal of Applied Genetics,* **47,** 277–286.

Gertz, J. M., Vencill, W. K. & Hill, N. S. (1999) Tolerance of transgenic soybean (Glycine max) to beat stress. *1999 Brighton Conference: Weeds,* **1–3,** 835–840.

Gottschling, D. E., Aparicio, O. M., Billington, B. L. & Zakian, V. A. (1990) Position effect at Saccharomyces-cerevisiae telomeres – reversible repression of Pol-Ii transcription. *Cell,* **63,** 751–762.

Griffey, C. A. & Allan, R. E. (1986) Effectiveness of stripe rust resistance among Lemhi 53 spring wheat near-Isogenic lines. *Crop Science,* **26,** 489–493.

Haldane, J. B. S. (1927) A mathematical theory of natural and artificial selection. Part III. *Proceedings of the Cambridge Philosophical Society,* **23,** 363–372.

Henikoff, S. (1979) Position effects and variegation enhancers in an autosomal region of *Drosophila melanogaster*. *Genetics,* **93,** 105–115.

Hill, R. A. & Sendashonga, C. (2003) General principles for risk assessment of living modified organisms: lessons from chemical risk assessment. *Environmental Biosafety Research,* **2,** 81–8.

Jones, H. D. (2005) Wheat transformation: current technology and applications to grain development and composition. *Journal of Cereal Science,* **41,** 137–147.

Joshi, J., Schmid, B., Caldeira, M. C., Dimitrakopoulos, P. G., Good, J., Harris, R., Hector, A., Huss-Danell, K., Jumpponen, A., Minns, A., Mulder, C. P. H., Pereira, J. S., Prinz, A., Scherer-Lorenzen, M., Siamantziouras, A. S. D., Terry, A. C., Troumbis, A. Y. & Lawton, J. H. (2001) Local adaptation enhances performance of common plant species. *Ecology Letters,* **4,** 536–544.

Kingsolver, J. G., Hoekstra, H. E., Hoekstra, J. M., Berrigan, D., Vignieri, S. N., Hill, C. E., Hoang, A., Gibert, P. & Beerli, P. (2001) The strength of phenotypic selection in natural populations. *American Naturalist,* **157,** 245–261.

Larkin, P. J. & Scowcroft, W. R. (1981) Somaclonal variation - a novel source of variability from cell-cultures for plant improvement. *Theoretical and Applied Genetics,* **60,** 197–214.

Last, F. T. (1953) Some effects of temperature and nitrogen supply on wheat powdery mildew. *Annals of Applied Biology,* **40,** 312–322.

Masci, S., D'Ovidio, R., Scossa, F., Patacchini, C., Lafiandra, D., Anderson, O. D. & Blechl, A. E. (2003) Production and characterization of a transgenic bread wheat line over-expressing a low-molecular-weight glutenin subunit gene. *Molecular Breeding,* **12,** 209–222.

McCullagh, P. J. & Nelder, J. (1989) *Generalised linear models.* Chapman and Hall, London.

Ortelli, S., Winzeler, H., Winzeler, M., Fried, P. M. & Nosberger, J. (1996) Leaf rust resistance gene Lr9 and winter wheat yield reduction .1. Yield and yield components. *Crop Science,* **36,** 1590–1595.

Pellegrineschi, A., Noguera, L. M., Skovmand, B., Brito, R. M., Velazquez, L., Salgado, M. M., Hernandez, R., Warburton, M. & Hoisington, D. (2002) Identification of highly transformable wheat genotypes for mass production of fertile transgenic plants. *Genome,* **45,** 421–430.

Peter, M., Lindfeld, A. & Nentwig, W. (2010) Does GM wheat affect saprophagous *Diptera* species (*Drosophilidae, Phoridae*)? *Pedobiologia,* **53,** 271–279.

Purrington, C. B. & Bergelson, J. (1995) Assessing weediness of transgenic crops - industry plays plant ecologist. *Trends in Ecology & Evolution,* **10,** 340–342.

Reed, J., Privalle, L., Powell, M. L., Meghji, M., Dawson, J., Dunder, E., Suttie, J., Wenck, A., Launis, K., Kramer, C., Chang, Y. F., Hansen, G. & Wright, M. (2001) Phosphomannose isomerase: An efficient selectable marker for plant transformation. *In Vitro Cellular & Developmental Biology-Plant,* **37,** 127–132.

Rooke, L., Byrne, D. & Salgueiro, S. (2000) Marker gene expression driven by the maize ubiquitin promoter in transgenic wheat. *Annals of Applied Biology,* **136,** 167–172.

Schlichting, C. D. (1986) The evolution of phenotypic plasticity in plants. *Annual Review of Ecology and Systematics,* **17,** 667–693.

Schmid, B. (1992) Phenotypic variation in plants. *Evolutionary Trends in Plants,* **6,** 45–60.

Slade, A. J., Fuerstenberg, S. I., Loeffler, D., Steine, M. N. & Facciotti, D. (2005) A reverse genetic, nontransgenic approach to wheat crop improvement by TILLING. *Nature Biotechnology,* **23,** 75–81.

Snow, A. A., Andow, D. A., Gepts, P., Hallerman, E. M., Power, A., Tiedje, J. M. & Wolfenbarger, L. L. (2005) Genetically engineered organisms and the environment: Current status and recommendations. *Ecological Applications,* **15,** 377–404.

Southern, E. (2006) Southern blotting. *Nature Protocols,* **1,** 518–525.

Stam, M., Viterbo, A., Mol, J. N. M. & Kooter, J. M. (1998) Position-dependent methylation and transcriptional silencing of transgenes in inverted T-DNA repeats: Implications for posttranscriptional silencing of homologous host genes in plants. *Molecular and Cellular Biology,* **18,** 6165–6177.

Stoger, E., Williams, S., Keen, D. & Christou, P. (1999) Constitutive versus seed specific expression in transgenic wheat: temporal and spatial control. *Transgenic Research,* **8,** 73–82.

Sultan, S. E. (1987) Evolutionary implications of phenotypic plasticity in plants. *Evolutionary Biology,* **21,** 127–178.

Sultan, S. E. (2001) Phenotypic plasticity for fitness components in *Polygonum* species of contrasting ecological breadth. *Ecology,* **82,** 328–343.

Tian, D., Traw, M. B., Chen, J. Q., Kreitman, M. & Bergelson, J. (2003) Fitness costs of R-gene-mediated resistance in *Arabidopsis thaliana. Nature,* **423,** 74–77.

Travella, S., Klimm, T. E. & Keller, B. (2006) RNA interference-based gene silencing as an efficient tool for functional genomics in hexaploid bread wheat. *Plant Physiology,* **142,** 6–20.

von Burg, S., Müller, C. B. & Romeis, J. (2010) Transgenic disease-resistant wheat does not affect the clonal performance of the aphid Metopolophium dirhodum Walker. *Basic and Applied Ecology,* **11,** 257–263.

Waines, J. G. & Hegde, S. G. (2003) Intraspecific gene flow in bread wheat as affected by reproductive biology and pollination ecology of wheat flowers. *Crop Science,* **43,** 451–463.

Wolfenbarger, L. L. & Phifer, P. R. (2000) The ecological risks and benefits of genetically engineered plants. *Science,* **290,** 2088–2093.

Yahiaoui, N., Srichumpa, P., Dudler, R. & Keller, B. (2004) Genome analysis at different ploidy levels allows cloning of the powdery mildew resistance gene *Pm3b* from hexaploid wheat. *Plant Journal,* **37,** 528–538.

Yu, L., Niu, J. S., Chen, P. D., Ma, Z. Q. & Liu, D. J. (2005) Cloning, physical mapping and expression analysis of a wheat mlo-like. *Journal of Integrative Plant Biology,* **47,** 214–222.

Zadoks, J. C., Chang, T. T. & Konzak, C. F. (1974) Decimal code for growth stages of cereals. *Weed Research,* **14,** 415–421.

Figure 2: Photograph taken by Simon Zeller

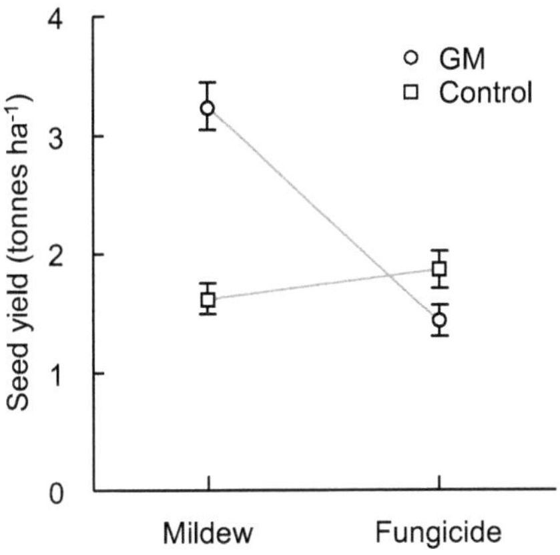

Figure 1. Effects of mildew infection and fungicide spraying on yields of GM wheat lines. Example of significant transgene × environment (presence/absence of powdery mildew) interaction in GM spring wheat in a glasshouse experiment. GM plants (circles = *Pm3b*#1 to #4) have higher yield than control plants (squares = S3b#1–4) in the presence but lower yield in the absence of mildew (fungicide spraying); light grey lines were drawn to make interactions between transgene and environments visible; error bars represent ± 1 standard error (back-transformed from square root scale).

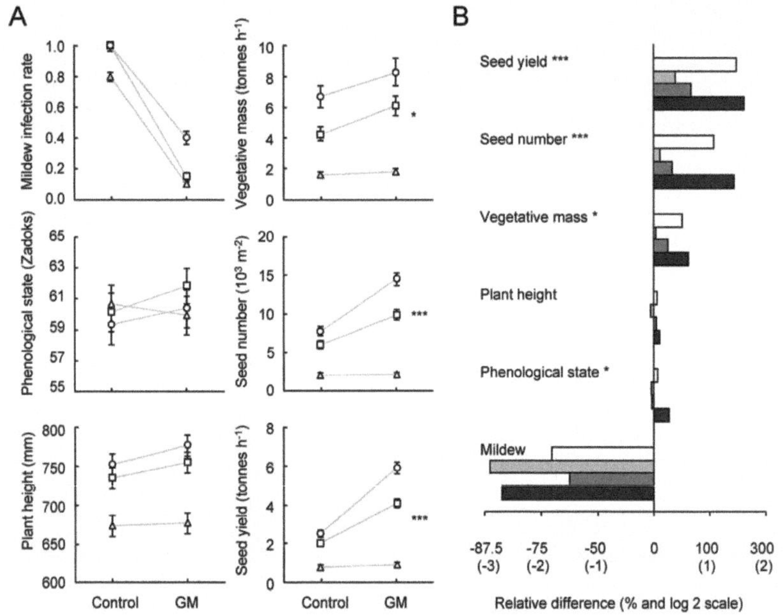

Figure 2. Effects of the transgene in the glasshouse on mildew infection and plant performance traits. The mildew infection equals the proportion of pots with strong powdery mildew infection up to flag leaves. Phenological stage, plant height, vegetative mass, seed number and seed yield were measured to assess the plant performance. A: mean of four lines (Control = S3b#1–4; GM = Pm3b#1–4) at different soil nutrient levels (circles = high fertilizer, squares = medium fertilizer, triangles = no additional fertilizer); significant transgene × fertilizer environment interactions indicated by asterisks (vegetative mass: P=0.035, seed number: P<0.001, seed yield: P<0.001); light grey lines were drawn to make these interactions visible; error bars represent ± 1 standard error (back-transformed, see methods) and are sometimes hidden behind the symbols. B: proportional difference between GM and control plants for each of the four offspring lines but averaged across nutrient levels (white bars = offspring pair 1 (Pm3b#1 *vs.* S3b#1), light grey = offspring pair 2, dark gray = offspring pair 3, black bars = offspring pair 4); x-axis log-scale with original values (100 * GM/control); bars extending to the right from the vertical zero line indicate higher values in GM than in control plants; significant GM/control x offspring pair interactions indicated by asterisks (* P<0.05; ***P<0.001).

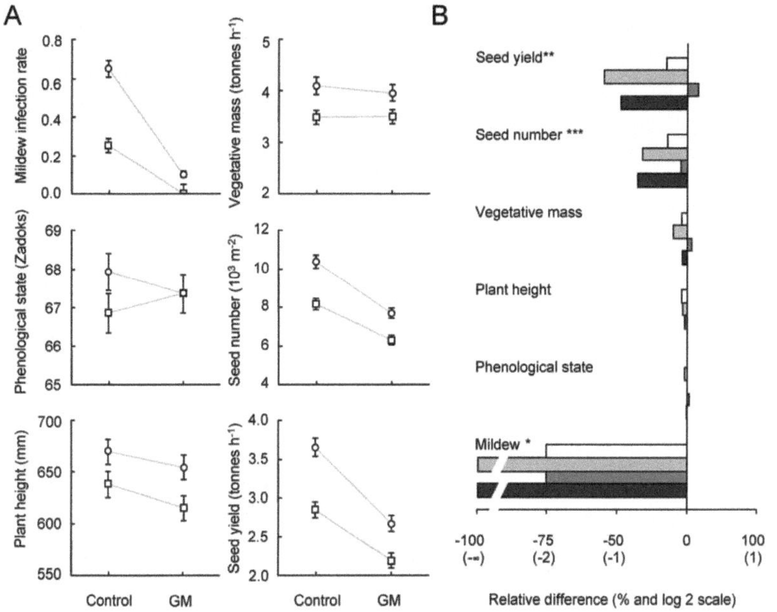

Figure 3. Effects of the transgene in the field on mildew infection and plant performance traits. The mildew infection equals the proportion of pots with strong powdery mildew infection up to flag leaves. Phenological stage, plant height, vegetative mass, seed number and seed yield were measured to assess the plant performance. A: mean of four lines at different soil nutrient levels (circles = additional fertilizer, squares = no fertilizer); transgene × fertilizer environment interactions were never significant; light grey lines were drawn to make this visible; error bars represent ± 1 standard error (back-transformed, see methods). B: proportional difference between GM and control plants for each of the four offspring lines but averaged across nutrient levels (white bars = offspring pair 1 (Pm3b#1 *vs.* S3b#1), light grey = offspring pair 2, dark gray = offspring pair 3, black bars = offspring pair 4); x-axis log-scale with original values (100 * GM/control); bars extending to the right from the vertical zero line indicate higher values in GM than in control plants; significant GM/control x offspring pair interactions indicated by asterisks (* $P<0.05$; ** $P<0.01$; *** $P<0.001$).

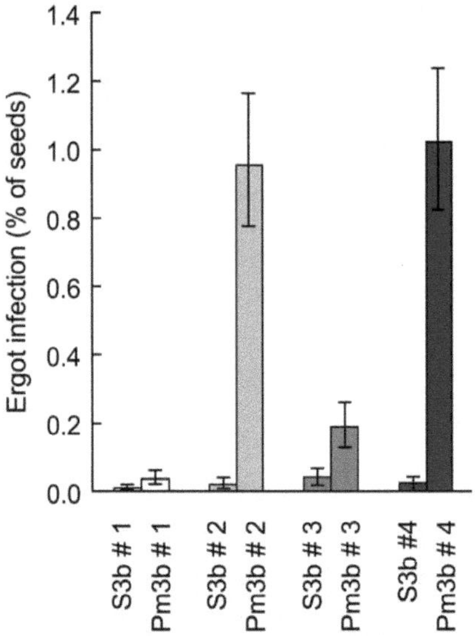

Figure 4. Percentage of ergot infected seeds in GM and control plants in the field. White bars = offspring pair 1, light grey = offspring pair 2, dark grey = offspring pair 3, black bars = offspring pair 4. Within each pair, the bar to the left shows control line and the bar to the right shows the corresponding GM line. Error bars represent ± 1 standard error (back-transformed from cube root scale.

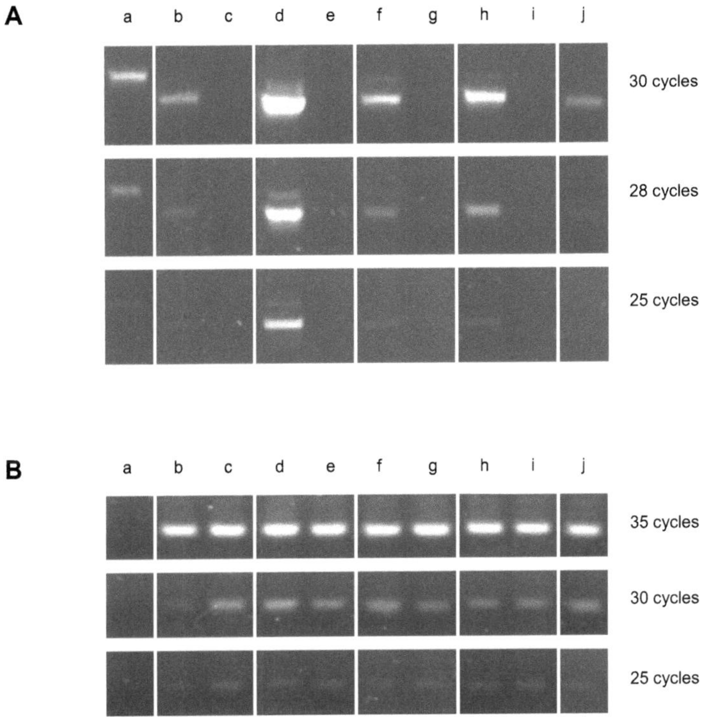

Figure S1. Semiquantitative expression analysis of *Pm3b* and *Mlo* in GM wheat lines. These gel photographs show semi-quantitative PCR expression analyses. A: Analysis of *Pm3b* expression in the *T. aestivum* lines *Pm3b* #1–4 (b, d, f, h) and the corresponding control lines S3b #1–4 (c, e, g, i). As positive controls, genomic DNA (a) and cDNA (j) of the variety Chul carrying one endogenous copy of *Pm3b* were used. The number of PCR cycles is indicated on the right. The photographs of the gel were cropped and rearranged graphically. B: As control for equal amount and quality of template cDNA, the expression levels of the *Mlo* gene were determined. Negative control water (a), *Pm3b* #1–4 (b, d, f, h), corresponding control lines S3b #1–4 (c, e, g, i), variety Chul (j).

Table S1. This ANOVA table shows the effect of the Fertilizer, GM / control, Offspring pair treatments and their interactions on the phenological state, plant height, vegetative mass, seed number and seed yield in the glasshouse experiment.

		Phenological state		Plant height		Vegetative mass		Seed number		Seed yield	
Source of variation	df	% SS	F pr.	% SS	F pr.	% SS	F pr.	% SS	F pr.	% SS	F pr.
Block	4	7.1	0.018	0.4	0.933	0.2	0.534	0.7	0.374	0.6	0.379
Fertilizer	2	0.5	0.651	26.3	<.001	86.1	<.001	61.8	<.001	55.6	<.001
GM / control	1	0.3	0.500	1.2	0.142	2.8	<.001	8.2	<.001	16.3	<.001
Offspring pair	3	26.4	<.001	15.8	<.001	1.0	0.006	2.5	0.004	1.6	0.014
GM / control x Offspring pair	3	6.0	0.017	3.4	0.099	0.9	0.012	4.7	<.001	3.9	<.001
Fertilizer x GM / control	2	0.6	0.586	0.4	0.720	0.5	0.035	4.1	<.001	6.8	<.001
Fertilizer x Offspring pair	6	3.5	0.410	0.6	0.979	0.5	0.394	0.9	0.536	0.5	0.717
Fertilizer x GM / control x Offspring pair	6	3.8	0.350	3.0	0.479	0.6	0.286	1.2	0.319	1.3	0.197
Residual	92	51.8		49.0		7.3		15.8		13.3	
Total	119	100.0		100.0		100.0		100.0		100.0	
		x^4 transformed				Sqrt transformed		Sqrt transformed		Sqrt transformed	
								1 plant without seeds excluded		1 plant without seeds excluded	

Table S2. This ANOVA table shows the effect of the Fertilizer, GM / control, Offspring pair treatments and their interactions on the phenological state, plant height, vegetative mass, seed number, seed yield and ergot infected seeds in the field experiment.

		Phenological state		Plant height		Vegetative mass		Seed number		Seed yield		Ergot Infection	
Source of variation	df	% SS	F pr.	% SS	F pr.	% SS	F pr.	% SS	F pr.	% SS	F pr.	% SS	F pr.
Block	3	69.1	<.001	3.0	0.583	3.7	0.511	3.7	0.073	4.9	0.035	2.1	0.229
Fertilizer	1	0.6	0.288	12.7	0.006	15.2	0.003	20.0	<.001	19.4	<.001	0.1	0.719
GM / control	1	0.0	0.935	3.7	0.126	0.1	0.761	32.3	<.001	31.7	<.001	40.3	<.001
Offspring pair	3	3.7	0.083	1.8	0.759	3.5	0.538	7.5	0.004	3.5	0.097	20.0	<.001
GM / control x Offspring pair	3	1.5	0.418	1.7	0.768	3.1	0.592	11.9	<.001	14.1	<.001	14.8	<.001
Fertilizer x GM / control	1	0.6	0.288	0.1	0.754	0.2	0.702	0.6	0.271	0.6	0.284	0.3	0.415
Fertilizer x Offspring pair	3	0.7	0.720	6.6	0.239	1.6	0.806	0.3	0.908	0.4	0.852	0.8	0.625
Fertilizer x GM / control x Offspring pair	3	0.7	0.720	2.4	0.665	0.7	0.936	1.5	0.397	1.7	0.364	0.3	0.884
Residual	45	23.2		68.0		71.8		22.2		23.6		21.3	
Total	63	100.0		100.0		100.0		100.0		100.0		100.0	
		x^4 transformed				Square root transformed		Sqare root transformed		Square root transformed		Cube root transformed	

Table S3. These ANOVA tables shows the effect of the Fertilizer, GM / control, Offspring pair treatments and their interactions (3 way interaction omitted) on the mildew infection rate in the glasshouse and field experiment.

	Glasshouse			Field		
Source of variation	df	% SS	F pr.	df	% SS	F pr.
Block	4	0.5	0.854	3	2.1	0.413
Fertilizer	2	3.2	0.016	1	9.8	<.001
GM / control	1	48.4	<.001	1	32.0	<.001
Offspring pair	3	1.8	0.186	3	6.1	0.051
GM / control x Offspring pair	3	2.0	0.15	3	7.8	0.022
Fertilizer x GM / control	2	2.3	0.052	1	0.3	0.538
Fertilizer x Offspring pair	6	5.4	0.032	3	6.6	0.039
Residual	98	36.4		48	35.2	
Total	119	100.0		63	100.0	

CHAPTER 2

Costs of resistance in genetically modified wheat

S.L. Zeller, O. Kalinina, B. Schmid, *Manuscript*

Fig. 3: Microplot experiment with inserted phytometer islands in April 2009

Abstract

Many resistance genes show costs of resistance. These costs are preferably measured in the absence of pathogens using genetically modified (GM) plants that differ in only one or a few resistance genes from control plants. To assess the ecological relevance of such costs, we grew individual plants of four transgenic spring wheat lines in a field trial with three pathogen levels and varied the genetic diversity of the crop.

We found that two lines with a *Pm3b* transgene were more resistant to powdery mildew than their sister lines of the variety Bobwhite whereas lines with *chitinase* or *chitinase* and *glucanase* transgenes were not more resistant than their mother variety Frisal. In the absence of the pathogen, all four GM lines performed worse than their controls, i.e. *Pm3b*#1 and *Pm3b*#2 had 46% and A9 and A13 had 18% lower yields. In the presence of the pathogen, all GM lines except *Pm3b*#2 could increase their yields and other fitness-related traits, reaching the performance levels of the control lines. Line *Pm3b*#2 seemed to have lost its phenotypic plasticity and had low performance in all environments. This may have been caused by very high transgene expression. No synergistic effects of mixing different GM lines with each other were detected. This might have been due to high transgene expression or the similarity between the lines regarding their resistance genes.

We conclude that costs of resistance can be high for transgenic plants with constitutive transgene expression even in cases where they are not more resistant than control lines. Transgenic plants could only compete with conventional varieties in environments with high pathogen pressure. Furthermore, the large variability among the GM lines which may be due to unpredictable transgene expression suggests that case-by-case assessments are necessary to evaluate costs of resistance.

Introduction

Plants interact with their environment in various ways. They have to compete with their neighbours and endure abiotic stresses and pathogen attacks. Natural selection can improve competitiveness and stress resistance. However, there are no wild plants with resistances against all possible pathogens (Bergelson and Purrington 1996). There seems to be a trade-off between performance and defence (Herms and Mattson 1992). Hence, genes that increase resistance against pathogens should be costly for a plant. A meta-analysis showed that resistant plants had lower fitness than non-resistant ones in approximately half of 88 studies considered (Bergelson and Purrington 1996). It is important to understand the mechanism leading to such costs and how these affect plant–pathogen systems. Such knowledge can be relevant for basic ecology as well as for agricultural ecosystems (Brown 2002).

Fitness costs that are associated with pathogen resistance are difficult to measure. Resistance genes are often linked to other genes making it almost impossible to elucidate single-gene costs of resistance. This problem can be avoided if transgenic (genetically modified = GM) plants that differ only in one or a few known genes from their original genetic background (Purrington 2000; Burdon and Thrall 2003) are used. Advances in crop engineering resulted in an enormous variety of transgenic plants that ecologists can use as model organisms.

There have been several studies that measured costs of resistance in transgenic plants (Purrington 2000; Burdon and Thrall 2003; Tian et al. 2003; Romeis et al. 2007). Resistance costs were found to be associated with many but not all transgenic plants (Snow et al. 1999). However, even if such costs exist, they have to be put into the right context. There are very few studies that varied the pathogen pressure which is necessary to study the ecological relevance of costs associated with resistance genes. The pathogen level can itself be influenced by the plant community which can either facilitate or slow down the spread of epidemics. Thus, studies have shown that the genetic diversity of a plant stand can reduce the pathogen pressure and therefore increase the performance at the level of the population and of individual plants (Schmid 1994, Wolfe 2000, Mundt 2002). However, we did not find any published reports where the influence of pathogen pressure and community diversity on plant performance and costs of resistance were evaluated in combination.

We therefore performed a field trial with four transgenic and two non-transgenic lines of spring wheat *Triticum aestivum* L. that belonged either to the variety Bobwhite or Frisal. The GM Bobwhite lines *Pm3b*#1 and *Pm3b*#2 harboured a *Pm3b* transgene against powdery mildew *Blumeria graminis* f.sp. *tritici* (DC.) Speer, whereas the Frisal lines A9 and A13 had either a *chitinase* or a *chitinase* and a *glucanase* gene to induce quantitative fungal resistance. These transgenic lines were produced from commercially available Bobwhite or Frisal plants which we took as controls. We established three fungal infection treatment levels. One third of the study plants were sprayed with fungicide to prevent powdery mildew infection. This allowed us to measure potential costs of resistance in the absence of the pathogen. Furthermore, plants were naturally or artificially infected with powdery mildew to obtain different pathogen infection levels. We worked with individual plants that were hand-seeded into plots containing either Bobwhite or Frisal lines of varying genetic diversity (0, 1 or 2 GM lines). The factorial fungal infection treatment x genetic diversity design allowed us to address the following questions: (i) are there differences between GM and non-GM lines and between different GM lines? (ii) are there costs of resistance in the absence of pathogens? And (iii) does the mixing of plant lines and therefore the increasing of genetic diversity increase resistance and performance and are there interactions between fungal infection and diversity treatments?

Materials and methods

Genetically modified wheat

We used six spring wheat lines of the Mexican variety Bobwhite SH 98 26 (Peter et al. 2010; von Burg et al. 2010; Zeller et al. 2010; Lindfeld et al. 2011; Brunner et al. 2011; von Burg et al. 2011) and the Swiss variety Frisal (Bieri et al. 2003) for our experiment. Two GM and one non-GM line were chosen from each variety. The GM lines of Bobwhite harboured a *Pm3b* transgene in different position on the genome, each derived from different transformation events. *Pm3b* confers race-specific resistance to powdery mildew and was obtained from the hexaploid wheat variety Chul (Yahiaoui et al. 2004). The lines which were named *Pm3b#1* and *Pm3b#2* were generated by biolistic transformation (Pellegrineschi et al. 2002). The plasmids pAHC17+NotI (*PMI*) and pAHC17+3NotI (*Pm3b*) were used as vectors (Christensen and Quail 1996; Travella et al. 2006). After *Not*I (for *Pm3b*) or *NotI*/*Hin*dIII (for *PMI*) digestion, only the desired fragments, but no vector sequences, were co-bombarded into wheat. The *Pm3b* gene was cloned under the control of the *Zea mays* L. (maize) ubiquitin promoter (Christensen and Quail 1996) and transformants were selected on mannose-containing media using the phosphomannose isomerase (PMI)-coding gene as selectable marker (Reed et al. 2001). Presence of the transgenes was confirmed by Southern hybridization analysis (Southern 2006) using probes from the PM3B- (bp 1231-1956 as referred to in the GenBank accession AY325736) and PMI- (bp 271 - 810 as referred to in the GenBank accession AAC74685) encoding regions. The GM lines contained the *Pmi* gene as well as one complete copy of *Pm3b*, which segregated as a single Mendelian locus in the T1 generation. Two *Pm3b* lines were multiplied to T5 and used for the field experiment. The level of transgene expression was assessed by quantitative RT-PCR using RNA isolated from leaves of field-grown plants. It revealed that *Pm3b* genes in the lines *Pm3b*#1 and *Pm3b*#2 were on average 11 and 55 times higher expressed than in the variety Chul where this gene is expressed naturally (Brunner et al. 2011).

The two transgenic lines with the genetic background of the variety Frisal contained genes from barley which are known for their anti-fungal effect and the constitutive or inducible expression of pathogenesis-related genes (Zhu et al. 1994). Line A9 harboured a *chitinase* and A13 both a *chitinase* and a β-1,3-*glucanase* transgene (Bliffeld et al. 1999). Both lines were generated by biolistic transformation (Pellegrineschi et al. 2002). A maize ubiquitin promoter (Christensen and Quail 1996) was used for the *chitinase* and an actin promoter from rice (McElroy et al. 1990) for the β-1,3-*glucanase*. Furthermore, a *Bar* coding sequence with a CaMV 35S promoter was introduced for selection. The *Bar* gene provides resistance against the herbicide glufosinate (Thompson et al. 1987). Analysis for *Bar* expression identified lines with clear 3:1 segregation in the second generation, indicating a single transgene integration locus. The expression of the transgenes *chitinase* and β-1,3-*glucanase* was analyzed by SDS-PAGE and Western blotting of intercellular wash fluid from mature leaves and in later generations on total protein from seedling leaves. Fungal infection essays carried out in the laboratory showed that the susceptibility of primary leaves of line A9 was reduced by 30% compared to control lines. Line A13, however, was more susceptible (130%) to powdery mildew than the control lines. Bieri et al. (2003) argued that the high expression of *glucanase* might have influenced the fungal resistance negatively. No morphological differences were observed. Both lines were multiplied to T6 in the glasshouse in order to verify stable expression of the transgenes.

Field experiment

The field experiment took place at an agricultural research station in Zurich-Reckenholz, Switzerland, at 440 m above sea level. It started in March 2009 and lasted until beginning of August 2009. Three powdery-mildew treatment blocks, each with twelve 1.0 x 1.3 m plots, were sown with seeds of the six GM lines described above

(Supplementary Figure S1). Besides the monocultures, six plots with 50:50 mixtures consisting of *Pm3b*#1/Bobwhite control, *Pm3b*#2/Bobwhite control, *Pm3b*#1/*Pm3b*#2 as well as A9/Frisal control, A13/Frisal control, A9/A13 were sown to assess mixture effects. In each plot five rows with a distance of 20 cm between them were sown at a density of 400 seeds per m^2 using a Seedmatic system (Hege 90, Hege Maschinen, Eging am See, Germany). To assess the performance of individual plants it was essential to know the line identity of plants in mixture plots. We therefore added small rows consisting of 7 seeds (seed island) of known identity by hand right after the machine seeding. Monocultures received one and mixture plots two islands. We positioned these islands parallel to the second or forth row with at least 20 cm distance to the plot boundaries. After emergence, the machine sown seedlings parallel to the seed island were removed. The seed in each seed island had therefore the same competitive environment as had the machine-sown seeds. Three out of the seven planted seeds (position 2, 4, 6) were marked with a label.

The three fungal infection treatments were fungicide application and natural and artificial mildew infection. Fungicide plots were sprayed three times with the fungicide Prosper (500g l^{-1} Spiroxamine; Leu + Gygax AG, Birmenstorf, Switzerland). This allowed keeping the plots almost completely free of powdery mildew. In the natural infection plots, neither artificial inoculation nor fungicides were applied. All untreated plots were infected strongly by powdery mildew during the field experiment. The plots with artificial powdery mildew infection were bordered with "spreader rows" of the susceptible conventional winter wheat variety Kanzler. The plants of the spreader rows had been pre-grown and inoculated with powdery mildew, isolate 96224, in the glasshouse. The distance between spreader rows and plots was 80 cm. The powdery mildew isolate 96224 had been collected between Winterthur and Kloten (Switzerland) in 1996 (Srichumpa et al. 2005; Brunner et al. 2010) and was known to be avirulent on *Pm3b* (Yahiaoui et al. 2009). A second batch of inoculated plantlets were produced and planted one month later. The three fungal infection treatments were separated from each other by a 4-m wide border crop of spring triticale to reduce cross-contamination.

Based on a nutrient assessment different amounts of nitrogen fertilizer were applied before sowing. This resulted in equal nitrogen concentrations (7.5g N m^{-2}) in each block. At the phenological stages 22–29 (Zadoks et al. 1974) additional nitrogen was added (3 g N m^{-2} as "Ammonsalpeter 27.5", Lonza, Visp, Switzerland). The natural field soil provided the plants with sufficient phosphorous, potassium and magnesium (81, 176 and 248 mg kg^{-1}). All plots were sprayed with the herbicide cocktail Concert

SX (40% Thifensulfurone, 4% Metusulfurone-methyl; Stähler Suisse AG, Zofingen, Switzerland) and Starane super (120 g l^{-1} Bromoxynil, 120 g l^{-1} Ioxynil, 100 g l^{-1} Fluroxypyr-metilheptil-ester; Omya Agro AG, Safenwil, Switzerland) in the beginning of May. Insecticide Karate Zeon (100g l^{-1} Lambda-Cyhalothrin; Syngenta Agro AG, Dielsdorf, Switzerland) against the wheat stem fly (*Chlorops pumilionis* Bjerk.) was applied in the beginning of May and repeated 2 weeks later.

Response variables

The degree of powdery mildew infection (Eyal et al. 1987) was assessed 32, 45, 59 and 80 days after germination. Based on these data, we calculated the “Area under Disease Progress Curve”, AUDPC (Jeger and Viljanen-Rollinson 2001). Furthermore, we measured the phenological stage (Zadoks et al. 1974) 59 days after germination. Leaf chlorophyll content was assed 80 days after germination with a chlorophyll meter (SPAD-502, Minolta Camera Co. Ltd, Japan) which measures light transmittance at red and near-infrared wavelength. The device provides SPAD values which are directly proportional to the total chlorophyll content. Plant height, defined as the highest point of the plant measured from the soil as well as spike length excluding awns were also recorded 80 days after germination. After ripening, all marked plants were cut at ground level and separated into vegetative and reproductive parts (spikes). At the same time, the number of spikes per plant was recorded. Vegetative and reproductive parts were then dried at 80 and 25 C°, respectively, and weighed. We then threshed the reproductive parts and obtained the seed mass which is equivalent to seed yield. Vegetative mass was calculated by subtracting the seed mass from the total biomass. Finally, the seed mass of the individual plants was divided by the number of seeds to calculate the thousand seed weight (TSW).

Data analysis

We analysed the data with mixed-model analysis of variance using the classical ANOVA as well as the REML (Restricted Maximum Likelihood) method with the statistical software GenStat (VSN International Ldt). Results were almost identical and thus only the REML analyses are presented in this chapter. They are summarized in tables for all variables (see supplementary Tables S1–S4). Fixed-effects terms were fitted with hierarchical and factorial models as follows. First, we used an “all hierarchical” treatment/line model that divided Bobwhite from Frisal plants and then analysed differences between GM and control lines, as well as between the two GM lines within each variety separately for each fungal infection treatment (Model 1; Figure S2a, Tables S1–S2). Second, we used a "factorial submodel" for fungal infection treatment x line within the two main groups Bobwhite and Frisal; in the submodel for each variety the main effects fungal infection and line and their interaction were fitted (Model 2; Figure S2b, Tables S3–S4).

Two additional terms were added to assess the influence on the target plants of the number of GM-lines (GM-richness 0, 1 and 2) or the proportion of GM-plants (GM-concentration 0, 50, 100%) per plot. Since these two contrasts were partly confounded with each other, their fitting sequence was alternated in two separate runs of the analyses. Furthermore, these contrasts were either fitted before or after the fungal infection treatment and line model (similar for line/fungal infection treatment model 1 and 2). Fitting the contrasts first in the models allowed an assessment of their influence "ignoring" line effects and fungal infection treatment x line interactions (fungal infection treatment main effects were not confounded with GM-richness or GM-concentration and therefore for this the fitting sequence did not matter). Fitting the contrasts after the fungal infection treatment and line effects allowed an assessment of their influence "eliminating" line effects and fungal infection treatment x line interactions, in other words, it allowed looking at effects of GM-richness and GM-concentration within groups defined by fungal infection treatment x line combinations (see e.g. McCullagh and Nelder 1989 for the ignoring/eliminating terminology). For example, significant GM-richness effects eliminating line effects therefore indicated that plants in plots with two GM lines behaved differently from plants in plots with one GM line — independently of the line identity — because differences between plots without or with GM lines had already been explained by the line effects.

To understand better the effects of fungal infection treatments and GM-richness and GM-concentration within each, Bobwhite or Frisal, we repeated all analyses with datasets restricted to either of the two varieties. Nevertheless, we mostly present results from the full model.

Residual plots were examined to check if the assumptions of normality and homoscedasticity were fulfilled. Seed yield, vegetative mass and seed number were square-root transformed and x^2 transformation was necessary for phenological state, plant height, spike length and TSW. Backtransformed means and standard errors from the REML output were used to draw the figures. The critical significance level was 0.05 in all analyses.

Since several of the measured traits correlated with each other, we also performed a Multivariate Linear Mixed Model (MLMM) to test for the overall significance of fungal infection treatment and line effects. The nine traits AUDPC, phenological state, chlorophyll content, plant height, spike length, seed yield, vegetative mass, spike number and seed number were combined in a single analysis. TSW was excluded because it was a linear combination of seed yield and seed number. Transformed data were used for the MLMM analysis.

Results

The spring wheat variety Bobwhite was more susceptible to powdery mildew than the old Swiss variety Frisal (Bobwhite vs. Frisal: $P<0.001$; Figure 1A and supplementary Table S1).The repeated spraying with fungicide reduced mildew infections by a factor of 6.2 for Bobwhite and by a factor of 5.4 for Frisal plants (Fungicide vs. mildew within Bobwhite or within Frisal both $P<0.001$, see supplementary Table S1). The natural and artificial mildew treatment levels did not differ significantly from each other with regard to mildew infection, both within Bobwhite or within Frisal. Nevertheless, we assume that the composition of the pathogen community differed between these two treatment levels because of the artificial infection with only one particular powdery mildew strain. The Bobwhite GM lines *Pm3b*#1 and *Pm3b*#2 were less susceptible to powdery mildew than the non-transgenic Bobwhite control line in all three fungal infection treatments (83, 52 and 61% less mildew in fungicide-treated, natural infection and artificial infection plots, respectively). *Pm3b*#2 had 36% less powdery mildew than *Pm3b*#1 in the plots with natural infection ($P<0.001$; supplementary Table S1). There was no such difference between the two Bobwhite GM lines in the plots with artificial infection were a mildew strain avirulent for *Pm3b* genes was released.

Mildew infections decreased with increasing GM-concentration and GM-richness in the plots ($P<0.001$ for GM-concentration and GM-richness fitted before line effects, data not shown). To understand why GM-concentration and GM-richness reduced the mildew infection levels in diverse plots, we performed further analyses. We fitted GM-concentration and GM-richness after fungal infection treatment and line effects and interactions and therefore eliminated these (see Material and methods). As a result, the significant results from above disappeared (see supplementary Tables S1 and S2) which means that the decreased powdery mildew infection can be explained by the different pathogen resistance levels of the individual lines. The GM-Frisal Lines A9 and A13 showed no increased pathogen resistance when compared to plants of the Frisal control line and also no differences for GM-concentration or GM-richness. The mixing of lines *Pm3b*#1 with *Pm3b*#2 or A9 with A13 did therefore not lead to synergistic reduction of powdery mildew infection levels.

Fungal infection treatment effects and differences between GM and control lines in these (all hierarchical model)

The performance of Bobwhite and Frisal plants depended strongly on the fungicide or mildew treatment levels and therefore on the pathogen pressure (separate MLMM for Bobwhite and Frisal both with P<0.001). Only Bobwhite but not Frisal lines performed differently in plots with natural as compared with artificial infection (MLMM for Bobwhite: P<0.001, MLMM for Frisal: P=0.496). Plants of the variety Bobwhite differed from Frisal in all traits except spike number (Figures 1 and 2; P<0.001 for all 9 traits, Figures 1 and 2, supplementary Tables S1–S2). We describe the Bobwhite results first, followed by Frisal.

The fungicide application slowed down the development and increased plant height and TSW within the Bobwhite variety (phenological state: P=0.004, plant height: P=0.009, TSW: P<0.001; supplementary Tables S1–S2). However, there were no overall positive effects on vegetative mass and seed yield because of line-specific responses to the fungicide application. Seed yields of plants of the Bobwhite control line and the GM line *Pm3b*#2 were 31% and 13% higher whereas they were 28% lower for plants of the GM line *Pm3b*#1.

Significant interactions GM vs. control x fungicide vs. no fungicide within variety Bobwhite were found for most traits (plant height: P=0.026, spike length: P=0.041, seed yield: P=0.002, seed number: P=0.024; supplementary Tables S3–S4). Comparing Bobwhite control and Bobwhite GM lines in the fungicide-treated plots the latter had 7% lower plant height, 14% shorter spikes, 46% lower seed yield, 34% lower vegetative mass, 19% fewer spikes and 42% fewer seeds than the control (P=0.0017 and P<0.001 for the other traits; supplementary Tables S1–S2). These results indicate that the Bobwhite GM lines, in contrast to the control line, did not benefit from the absence of the pathogens. Plants in the plots with artificial infection developed faster, had slightly higher chlorophyll content but less spikes than plants in plots with natural infection (phenological state: P<0.001, chlorophyll: P=0.042, spike number: P=0.015; supplementary Tables S1–S2). The hierarchical REML analysis (supplementary Tables S1–S2) revealed significant differences between GM and control lines in natural (chlorophyll: P=0.006, spike length: P<0.001, seed number: P=0.032, TSW: P=0.011) and artificial infection plots (Chlorophyll: $P=0.25$; spike length: $P<0.001$). However, in contrast to the fungicide treatment level data, there were additional significant differences between the two Bobwhite GM lines in natural and artificial infection plots.

Frisal lines that were sprayed with fungicide grew taller than unsprayed plants (Fungicide vs. Mildew within Frisal in supplementary Table 1, plant height: $P<0.001$). As for the Bobwhite lines, the Frisal GM lines had 6% lower plant height, 11% shorter spikes, 18% lower yield and 20% fewer seeds than the control line (contrast Frisal/GM in Fungicide; plant height and spike length: $P<0.001$, yield: $P=0.048$, seed number: $P=0.031$; supplementary Tables S1–S2) in the sprayed plots. No such differences were found for plants growing in plots with natural or artificial infection. Only spike length remained lower in all three treatment levels (Frisal/GM contrasts in Fungicide, Natural and Artificial infection treatment levels in supplementary Table S1: $P<0.001$, $P=0.007$ and $P=0.002$).

Differences between GM-lines (factorial submodel)

Although the two GM lines of Bobwhite, *Pm3b*#1 and *Pm3b*#2, had the same transgene, they had very different phenotypes ($P<0.001$; MLMM). *Pm3b*#2 had a 5% slower development, 21% less chlorophyll, a 5% reduced height, 7% shorter spikes, 41% lower seed yield, 19% lower vegetative mass, 19% fewer seeds and a 25% reduced TSW compared with *Pm3b*#1 (phenological stage, chlorophyll, plant height, spike length, seed yield and TSW all $P<0.001$, vegetative mass: $P=0.012$, seed number: $P=0.014$; supplementary Tables S3–S4). In addition to this overall difference, the two GM lines also showed different responses to the two mildew treatments levels (significant interaction fungicide vs. no fungicide x *Pm3b*#1 vs. *Pm3b*#2 for seed yield: $P=0.039$, vegetative mass: 0.010, seed number: 0.019, plant height: 0.041, spike length: $P=0.001$; supplementary Tables S3–S4). This was due to a higher relative performance of *Pm3b*#1 in plots with mildew than with fungicide whereas no such response was found for line *Pm3b*#2. However, even the GM line *Pm3b*#1 never reached the performance of control plants in fungicide plots. The yield of unsprayed *Pm3b*#1 was 21% and that of *Pm3b*#2 59% lower than that of the Bobwhite control line in the fungicide treatment level.

Also in the variety Frisal the two GM lines, A9 and A13, had different phenotypes ($P<0.001$, MLMM). Plants of line A9 were 4% shorter, had 5% shorter spikes and 8% lower TSW than A13 (spike length: $P=0.008$, plant height and TSW: $P<0.001$; supplementary Tables S3–S4). As for the Bobwhite GM lines, also the Frisal

GM lines could never reach the yields of sprayed Frisal control plants. Unsprayed A9 plants had 20% and unsprayed A13 plants had 27% lower seed yields than sprayed plants of the Frisal control line.

Effects of GM-concentration and GM-richness

The genetic diversity of the plot into which the tested plants were sown influenced their performance. Plots with higher GM-concentrations had lower chlorophyll content, plant height, seed yield, vegetative mass and seed number (chlorophyll, plant height, seed yield and seed number: $P<0.001$, vegetative mass: $P=0.022$). Plots with higher GM-richness harboured plants with higher TSW ($P<0.001$). To understand why GM-concentration had mostly negative effects on fitness-related traits, we fitted GM-concentration and GM-richness after line and fungal infection treatment effects and interactions and therefore eliminated these (see Materials and methods). With the exception of plant height, all significant results from above disappeared (see supplementary Tables S1 and S2). By looking at the data we could see that the good performance of Bobwhite control and the bad performance of line *Pm3b*#2 were responsible for most of the concentration and richness effects. No synergistic effects caused by the mixing of lines *Pm3b*#1 with *Pm3b*#2 or A9 with A13 were detected.

Discussion

Powdery mildew infection

Our results show that the two tested spring wheat varieties differed from each other. Bobwhite lines proved to be more susceptible to powdery mildew than the Swiss variety Frisal. This might have to do with different breeding aims and the origin of these varieties. In Switzerland, where powdery mildew is a serious plant disease, breeders have favoured resistant varieties whereas this was not necessary in Mexico where no natural epidemics occur (Lillemo et al. 2006). Nevertheless, the control lines of both Bobwhite and Frisal varieties were infected by this pathogen. The GM lines *Pm3b*#1 and *Pm3b*#2 proved to be more resistant to powdery mildew than their genetic background Bobwhite. No such differences were detected in the A9 and A13 lines which were produced from Frisal. This is in contrast to laboratory results were A9 was less susceptible to powdery mildew than Frisal (Bieri 2003). Hence, these results demonstrate the importance of field trials.

Since we worked in a natural environment it was not possible to remove the omnipresent natural mildew spores. However, the fungicide used in the fungicide treatment level reduced powdery mildew infections in all plots to almost zero. This allowed us to assess the influence of the pathogen pressure on fitness-related traits and unintended effects. The difference between the natural and artificial treatment levels was less prominent. There was no overall difference in pathogen abundance between these two treatment levels, although the artificial infection started before the natural infection. It is conceivable that climatic conditions and not the start of the inoculation mainly affected the spread and growth of powdery mildew. However, it is likely that the artificially introduced mildew isolate 96224 was more common in artificial than in natural infection plots. This strain is avirulent for the two Bobwhite GM lines *Pm3b*#1 and *Pm3b*#2. We therefore expected less mildew in these plots than in the naturally infected ones. Indeed, line *Pm3b*#1 proved to be more resistant in the artificially than in the naturally inoculated plots. Line *Pm3b*#2, however, was highly resistant in both and this could have been due to the very high transgene expression levels of this line that made it even resistant to a "non-target" powdery mildew strain. Brunner et al. (2011) argued that high expression does provide some degree of quantitative resistance against different strains of powdery mildew.

Besides the mildew treatment levels, we analysed the influence of plant diversity on individual plants within a plot. Plants in plots with high concentrations of resistant GM lines had less powdery mildew. This effect can be explained simply by the presence or absence of the susceptible Bobwhite line. One reason to include diversity treatments into our experimental design was to assess possible synergistic effects caused by the mixing of different GM lines. There are several publications that show improved pathogen resistance in fields with mixed varieties (Finckh et al. 2000; Wolfe 2000; Mundt 2002). However, we found no indications that mixed *Pm3b*#1 and *Pm3b*#2 plots were more resistant against powdery mildew than monocultures of these GM lines with identical transgenes but different expression levels. There are at least two explanations for this. Either the influence of the mixed background was not strong enough to affect the plants which themselves belonged to uniform seed islands or these lines were too similar to allow synergistic or complementary effects. The same might be true for the Frisal lines. Although not genetically identical, all three Frisal lines were similarly resistant against powdery mildew in all three fungal infection treatments. Hence, in the absence of variability, no synergistic effects can be expected.

Costs of Resistance

If a transgene would induce complete pathogen resistance without any costs we would expect GM lines to perform as well as non-resistant control lines in absence of the pathogen. We found however, that all four GM lines performed worse than their Bobwhite and Frisal control lines on fungicide-treated plots. Furthermore, none of the lines ever reached the level of the non-GM control lines in the un-sprayed plots. This indicates that *Pm3b* as well as *chitinase* and *glucanase* transgenes cause costs of resistance. We found that the disadvantage of GM lines decreased in plots with high pathogen levels. However, this is mainly due to lower performance of the control lines in the mildew-infected plots.

Whereas costs of resistance might explain why these GM lines did not reach the level of the control lines in the absence of the pathogen, this does not explain why line *Pm3b*#1 performed worse in the fungicide than in the mildew treatment levels. One explanation could be that the chemicals of the fungicide interacted with the transgene or its products. Increased sensitivity to fungicide was described already earlier in a glasshouse study (Zeller et al. 2010). The sum of costs of resistance and fungicide sensitivity could have caused the large fitness reductions in lines *Pm3b*#1 and *Pm2b*#2. Since it is not possible to remove a common pathogen from a field without the use of pesticide one would have to revert to closed systems without pathogen presence to study costs of resistance separate from potential fungicide effects. However, costs of resistance might not be visible under conditions that are optimal for plant growth. A better approach then closed systems might be to carry out field trials in areas where the targeted pathogen does not occur naturally, or to stress the plants in the closed system.

Whereas line *Pm3b*#1 performed better in the mildew treatment levels than in the fungicide treatment level presumably due to benefits related to its powdery mildew resistance; *Pm3b*#2 performed poorly in all environments. For this line, costs of resistance seemed to be so large that potential benefits of the transgene were offset. Thus, depending on the environment, line *Pm3b*#1 performed better, i.e. had retained more plasticity than line *Pm3b*#2. This difference might be explained by the expression level. Line *Pm3b*#2 is known for much higher transgene expression levels than line *Pm3b*#1 (Zeller et al. 2010; Brunner et al. 2011). It is conceivable that costs of resistance increase with higher expression level because of increased metabolic stress. Among the GM Frisal lines, A13 had higher plants, spike length and TSW. Seed yield and seed number were lower in line A13 but these differences were not significant. We could therefore not prove that line A13, which harbours two transgenes, performs worse than line A9 with only one. Further experiments are necessary to assess if the number of transgenes within a single plant increases costs of resistance (see also Chapter 4).

GM plants with high costs of resistance may not be particularly useful in agronomy. They have however one advantage: their risk of spreading uncontrollably in fields or even to natural habitats is very low. It is very likely that such plants would be outcompeted in natural habitats where pathogens are known to fluctuate widely.

Diversity effects

Besides the influence of the fungal infection treatments, we studied how the genetic diversity of plant communities influenced individual plants within these communities. There are examples from agronomy where increased diversity leads to reduced pathogen susceptibility and transgressive overyielding (Finckh et al. 2000; Wolfe 2000; Mundt 2002). If crop varieties or wild plant species are mixed with each other, it is difficult if not impossible test if particular resistance genes or other phenotypic traits are responsible for these positive diversity effects. Transgenic plants that differ only in single genes can be useful to understand such mechanisms. Hence, we planted either monocultures or mixtures of one GM with one non-GM line or two different GM lines. We found that chlorophyll content, plant height and several fitness-related traits were influenced by the concentration and TSW by the number of different GM plants within each plot. However, almost all of these differences could be explained by the presence of a particular line in a subset of plots. Hence, no benefits of mixing these GM lines with each other were detected. This is in line with the powdery mildew results which we discussed above. Individual plants were not less infected with this pathogen than expected from the monoculture means. The amount of powdery mildew infection seemed to influence the overall performance of our study plants. Thus, because powdery mildew was not reduced more in plots with two GM lines than in plots with only one we would also not expect positive effects on other traits. Furthermore, high costs of resistance might have concealed such effects. As described in the section above, the GM lines might have been too similar to complement each other, or the lack of mixing in the planted islands could have concealed the effects. We recommend, therefore, using more dissimilar transgenic plants for future diversity studies. Furthermore, better mixing might be necessary to obtain good diversity effects.

Conclusions

Our study demonstrates that transgenic plants may differ from their non-GM control lines in many traits and that these differences can be influenced by environmental factors (i). There were differences between the Bobwhite GM lines *Pm3b*#1 and *Pm3b*#2 as well as between the Frisal GM lines A9 and A13. The latter might be explained by differences in the introduced gene construct. The lines *Pm3b*#1 and *Pm3b*#2 share, however, an identical transgene. It is most likely that different expression levels caused by positional effects were responsible for the differences between the two Bobwhite GM lines. In view of all this variation, we conclude that ecological assessments of GM plants should be done on a case-by-case basis (Andow and Zwahlen 2006).

We found that all four tested GM lines suffered from costs of resistance in the absence of the pathogen (ii). Interestingly, even transgenic plants without increased pathogen resistance showed such negative effects. Three of the four tested GM lines did not differ in their performance from the non-GM control lines in presence of the pathogen. This does, however, not mean that costs of resistance were non-existent. It is more likely that positive effects of the pathogen resistance concealed costs of resistance. Finally, the diversity of the plant communities influenced pathogen levels and plant performance (iii). However, no synergistic effects were detected. We conclude that the balance between costs and benefits of increased pathogen resistance and therefore the performance of GM plants depends mainly on environmental factors. It is conceivable that transgenic plants with high costs of resistance can outperform conventional lines only in areas with constantly high pathogen pressure. Pathogen populations are known to vary from year to year depending mostly on weather conditions and other factors. Hence, in years of low pathogen pressure, non-resistant plants should have an advantage over resistant plants. One could therefore recommend to cultivate both resistant and non-resistant plants in places with variable pathogen populations.

Acknowledgments

We thank S. Brunner, B. Keller, C. Sautter, J. Fütterer and A. Fammartino for the seed material; The national research station Agroscope Reckenholz-Tänikon ART for setting up the field experiment and I. Kostetskyi and numerous helpers for assistance in the field. This project was supported by the Swiss National Science Foundation and is a part of the wheat-cluster.ch, a sub-unit of the national research programme NRP 59 (SNF 405940–115607).

References

Andow, D. A. & Zwahlen, C. (2006) Assessing environmental risks of transgenic plants. *Ecology Letters,* **9,** 196–214.

Bergelson, J. & Purrington, C. B. (1996) Surveying patterns in the cost of resistance in plants. *American Naturalist,* **148,** 536–558.

Bieri, S., Potrykus, I. & Futterer, J. (2003) Effects of combined expression of antifungal barley seed proteins in transgenic wheat on powdery mildew infection. *Molecular Breeding,* **11,** 37–48.

Bliffeld, M., Mundy, J., Potrykus, I. & Futterer, J. (1999) Genetic engineering of wheat for increased resistance to powdery mildew disease. *Theoretical and Applied Genetics,* **98,** 1079–1086.

Brown, J. K. M. (2002) Yield penalties of disease resistance in crops. *Current Opinion in Plant Biology,* **5,** 339–344.

Brunner, S., Hurni, S., Herren, G., Kalinina, O., von Burg, S., Zeller, S., Schmid, B., Winzeler, M. & Keller, B. (2011) Transgenic *Pm3b* wheat lines show resistance to powdery mildew in the field. *Plant Biotechnology Journal,* **In press**.

Brunner, S., Hurni, S., Streckeisen, P., Mayr, G., Albrecht, M., Yahiaoui, N. & Keller, B. (2010) Intragenic allele pyramiding combines different specificities of wheat *Pm3* resistance alleles. *Plant Journal,* **64,** 433–445.

Burdon, J. J. & Thrall, P. H. (2003) The fitness costs to plants of resistance to pathogens. *Genome Biology,* **4,** 227–229.

Christensen, A. H. & Quail, P. H. (1996) Ubiquitin promoter-based vectors for high-level expression of selectable and/or screenable marker genes in monocotyledonous plants. *Transgenic Research,* **5,** 213–218.

Eyal, Z., Scharen, A., Prescott, J. & van Ginkel, M. (1987) *The septoria diseases of wheat: concepts and methods of disease management.* International Maize and Wheat Improvement Center, D.F. Mexico.

Finckh, M. R., Gacek, E. S., Goyeau, H., Lannou, C., Merz, U., Mundt, C. C., Munk, L., Nadziak, J., Newton, A. C., de Vallavielle-Pope, C. & Wolfe, M. S. (2000) Cereal variety and species mixtures in practice, with emphasis on disease resistance. *Agronomy Journal,* **20,** 813–837.

Herms, D. A. & Mattson, W. J. (1992) The dilemma of plants – to grow or defend. *Quarterly Review of Biology,* **67,** 283–335.

Jeger, M. J. & Viljanen-Rollinson, S. L. H. (2001) The use of the area under the disease-progress curve (AUDPC) to assess quantitative disease resistance in crop cultivars. *Theoretical and Applied Genetics,* **102,** 32–40.

Lillemo, M., Skinnes, H., Singh, R. P. & van Ginkel, M. (2006) Genetic analysis of partial resistance to powdery mildew in bread wheat line Saar. *Plant Disease,* **90,** 225–228.

Lindfeld, A., Lang, C., Knop, E. & Nentwig, W. (2011) Hard to digest or a piece of cake? Does GM wheat affect survival and reproduction of *Enchytraeus albidus* (*Annelida: Enchytraeidae*)? *Applied Soil Ecology,* **47,** 51–58.

McCullagh, P and Nelder, J. A. (1989) *Generalized linear models.* Second Edition. Chapman and Hall, London.

McElroy, D., Zhang, W. G., Cao, J. & Wu, R. (1990) Isolation of an Efficient Actin Promoter for Use in Rice Transformation. *Plant Cell,* **2,** 163–171.

Mundt, C. C. (2002) Use of multiline cultivars and cultivar mixtures for disease management. *Annual Review of Phytopathology,* **40,** 381–410.

Pellegrineschi, A., Noguera, L. M., Skovmand, B., Brito, R. M., Velazquez, L., Salgado, M. M., Hernandez, R., Warburton, M. & Hoisington, D. (2002) Identification of highly transformable wheat genotypes for mass production of fertile transgenic plants. *Genome,* **45,** 421–430.

Peter, M., Lindfeld, A. & Nentwig, W. (2010) Does GM wheat affect saprophagous *Diptera* species (*Drosophilidae, Phoridae*)? *Pedobiologia,* **53,** 271–279.

Purrington, C. B. (2000) Costs of resistance. *Current Opinion in Plant Biology,* **3,** 305–308.

Reed, J., Privalle, L., Powell, M. L., Meghji, M., Dawson, J., Dunder, E., Suttie, J., Wenck, A., Launis, K., Kramer, C., Chang, Y. F., Hansen, G. & Wright, M. (2001)

Phosphomannose isomerase: an efficient selectable marker for plant transformation. *In Vitro Cellular & Developmental Biology-Plant,* **37,** 127–132.

Romeis, J., Waldburger, M., Streckeisen, P., Hogervorst, P. A. M., Keller, B., Winzeler, M. & Bigler, F. (2007) Performance of transgenic spring wheat plants and effects on non-target organisms under glasshouse and semi-field conditions. *Journal of Applied Entomology,* **131,** 593–602.

Schmid, B. (1994) Effects of genetic diversity in experimental stands of *Solidago altissima* - evidence for the potential role of pathogens as selective agents in plant-populations. *Journal of Ecology,* **82,** 165–175.

Snow, A. A., Andersen, B. & Jørgensen, R. B. (1999) Costs of transgenic herbicide resistance introgressed from *Brassica napus* into weedy *B. rapa. Molecular Ecology,* **8,** 605–615.

Southern, E. (2006) Southern blotting. *Nature Protocols,* **1,** 518–525.

Srichumpa, P., Brunner, S., Keller, B. & Yahiaoui, N. (2005) Allelic series of four powdery mildew resistance genes at the Pm3 locus in hexaploid bread wheat. *Plant Physiology,* **139,** 885–895.

Thompson, C. J., Movva, N. R., Tizard, R., Crameri, R., Davies, J. E., Lauwereys, M. & Botterman, J. (1987) Characterization of the herbicide-resistance gene bar from *Streptomyces hygroscopicus. Embo Journal,* **6,** 2519–2523.

Tian, D., Traw, M. B., Chen, J. Q., Kreitman, M. & Bergelson, J. (2003) Fitness costs of R-gene-mediated resistance in *Arabidopsis thaliana. Nature,* **423,** 74–77.

Travella, S., Klimm, T. E. & Keller, B. (2006) RNA interference-based gene silencing as an efficient tool for functional genomics in hexaploid bread wheat. *Plant Physiology,* **142,** 6–20.

von Burg, S., Müller, C. B. & Romeis, J. (2010) Transgenic disease-resistant wheat does not affect the clonal performance of the aphid *Metopolophium dirhodum Walker. Basic and Applied Ecology,* **11,** 257–263.

von Burg, S., van Veen, F. J. F., Álvarez-Alfageme, F. & Romeis, J. (2011) Aphid-parasitoid community structure on genetically modified wheat. *Biology Letters*, doi: 10.1098/rsbl.2010.1147.

Wolfe, M. S. (2000) Crop strength through diversity. *Nature,* **406,** 681–682.

Yahiaoui, N., Kaur, N. & Keller, B. (2009) Independent evolution of functional Pm3 resistance genes in wild tetraploid wheat and domesticated bread wheat. *Plant Journal,* **57,** 846–856.

Yahiaoui, N., Srichumpa, P., Dudler, R. & Keller, B. (2004) Genome analysis at different ploidy levels allows cloning of the powdery mildew resistance gene *Pm3b* from hexaploid wheat. *Plant Journal,* **37,** 528–538.

Zadoks, J. C., Chang, T. T. & Konzak, C. F. (1974) Decimal code for growth stages of cereals. *Weed Research,* **14,** 415–421.

Zeller, S. L., Kalinina, O., Brunner, S., Keller, B. & Schmid, B. (2010) Transgene × environment interactions in genetically modified wheat. *PloS ONE,* **5,** e11405.

Zhu, Q., Maher, E. A., Masoud, S., Dixon, R. A. & Lamb, C. J. (1994) Enhanced protection against fungal attack by constitutive coexpression of *chitinase* and *glucanase* genes in transgenic tobacco. *Bio-Technology,* **12,** 807–812.

Figure 3: Photograph taken by Simon Zeller

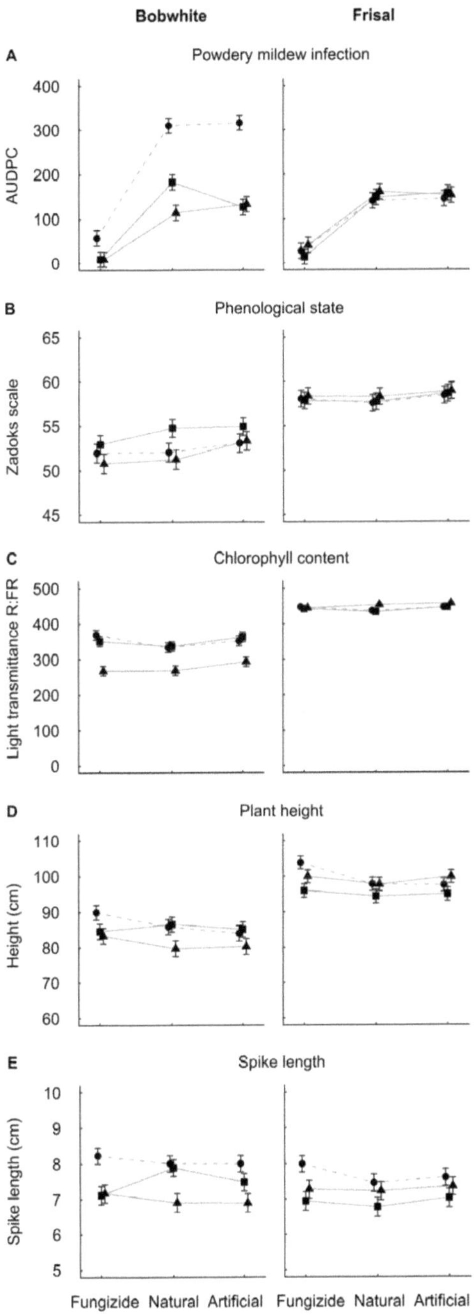
Bobwhite
Frisal
A
Powdery mildew infection
AUDPC
400
300
200
100
0
B
Phenological state
Zadoks scale
65
60
55
50
45
C
Chlorophyll content
Light transmittance R:FR
500
400
300
200
100
0
D
Plant height
Height (cm)
110
100
90
80
70
60
E
Spike length
Spike length (cm)
10
9
8
7
6
5
Fungizide Natural Artificial
Fungizide Natural Artificial

Figure 1. Effects of fungicide and natural and artificial powdery mildew infection on performance of GM and non GM-wheat. The left column shows the non-transgenic variety Bobwhite (dashed line, round symbols) and two transgenic lines *Pm3b#*1 (solid lines, square symbols) and *Pm3b#2* (solid lines, triangular symbols). The right column shows the non-transgenic variety Frisal (dashed line, round symbols) and two transgenic lines A9 (solid lines, square symbols) and A13 (solid lines, triangular symbols). A–E present the level of powdery mildew infection, phenological stage, chlorophyll content, plant height and spike length data. Light grey lines were drawn to make transgene x fungal infection treatment interactions visible; error bars represent ± 1 standard error (back-transformed, see Material and methods) and are sometimes hidden behind the symbols.

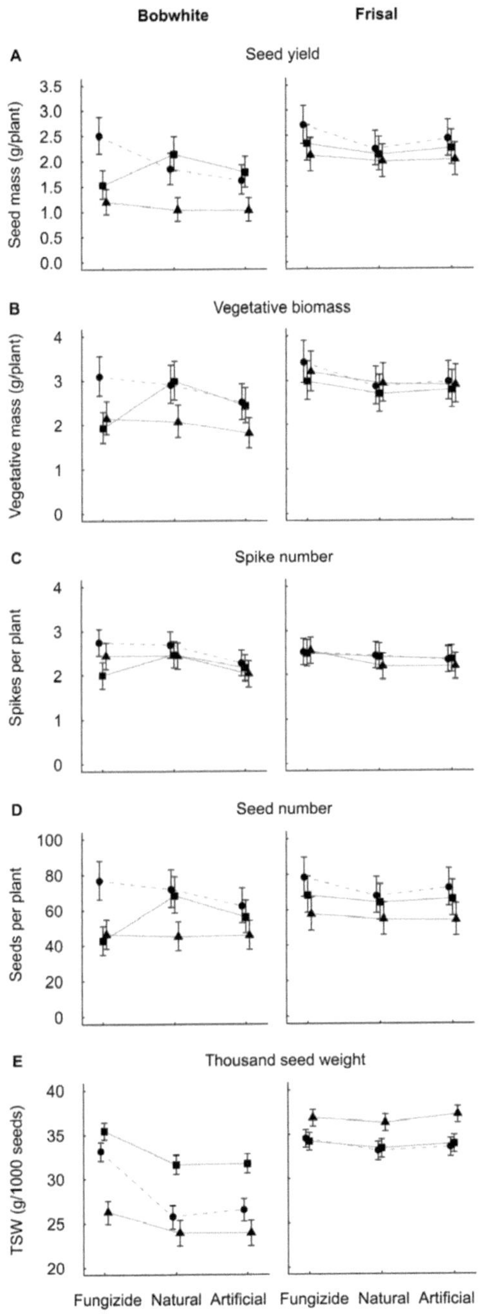
Bobwhite
Frisal
A
Seed yield
Seed mass (g/plant)
B
Vegetative biomass
Vegetative mass (g/plant)
C
Spike number
Spikes per plant
D
Seed number
Seeds per plant
E
Thousand seed weight
TSW (g/1000 seeds)
Fungizide
Natural
Artificial

Figure 2. Effects of fungicide and natural and artificial powdery mildew infection on performance of GM and non-GM wheat. The left column shows the non-transgenic variety Bobwhite (dashed line, round symbols) and two transgenic lines *Pm3b*#1 (solid lines, square symbols) and *Pm3b#2* (solid lines, triangular symbols). The right column shows the non-transgenic variety Frisal (dashed line, round symbols) and two transgenic lines A9 (solid lines, square symbols) and A13 (solid lines, triangular symbols). A–E present seed yield, vegetative biomass, spike number, seed number and thousand seed weight data. Light grey lines were drawn to make transgene x fungal infection treatment interactions visible; error bars represent ± 1 standard error (back-transformed, see Material and methods) and are sometimes hidden behind the symbols.

Figure S1. Experimental design

A. Block and treatment structure

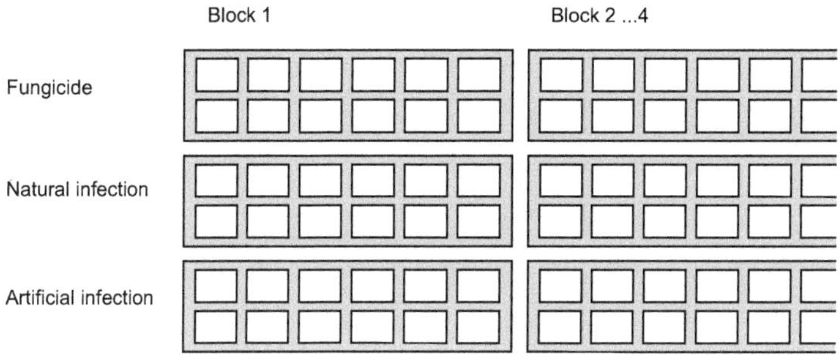

B. Plots in fungal infection treatment (example artificial infection)

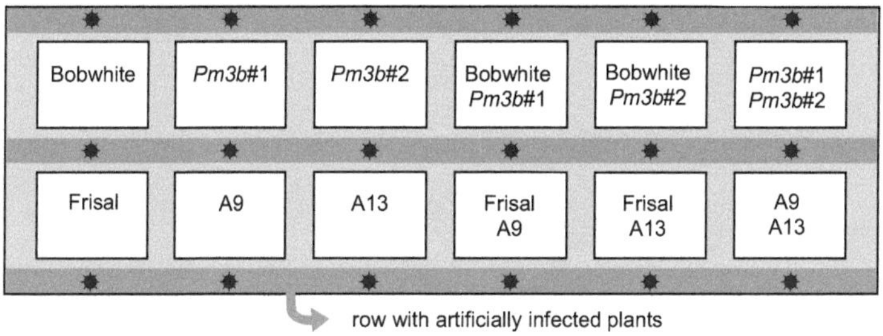

C. Seed islands in plot (example Bobwhite / *Pm3b*#1 mixture)

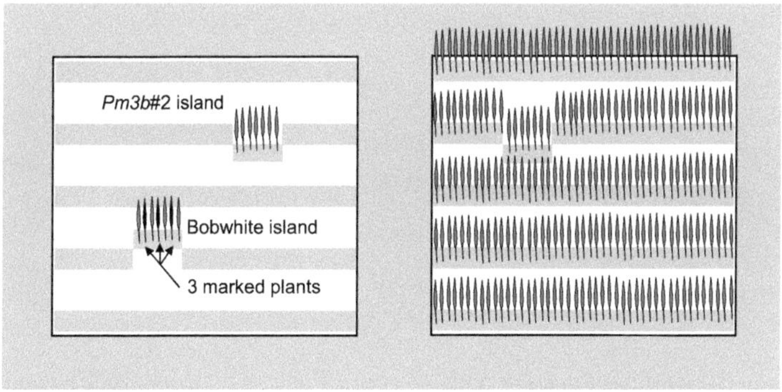

Figure S2. Fungal infection treatment/line models used in the analysis

A. Model 1 (all hierarchical)

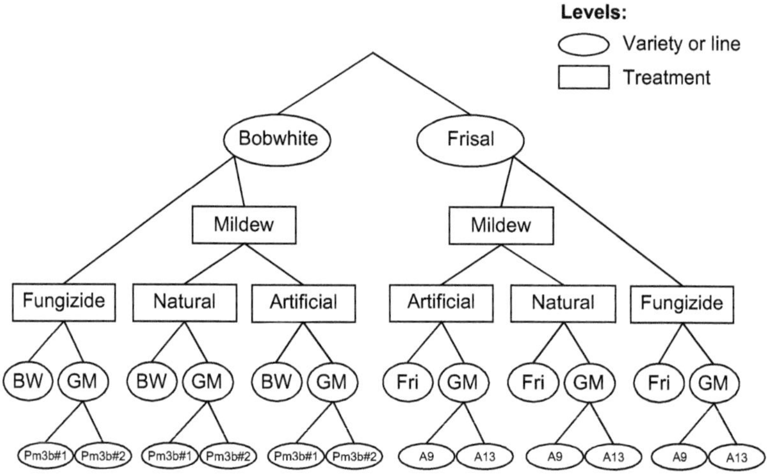

B. Model 2 (factorial submodel within Bobwhite and Frisal)

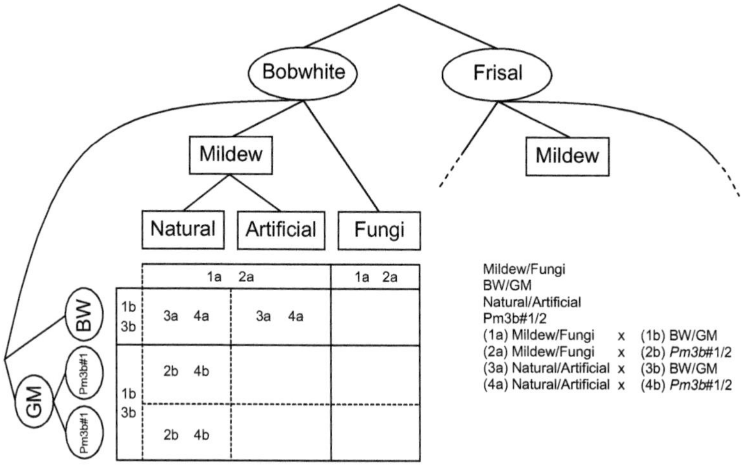

Table S1. Summary of REML tables of AUDPC, phenological state, chlorophyll content, plant height and spike length. A hierarchical model was used to test the fungal infection treatment and line effects. GM richness and GM concentration were alternated. Random terms are not included in the table since their variance components are estimated directly in the REML analyses. The percentage of Wald statistic thus was calculated only for the total of the fixed effects. The total is smaller than 100% because complex interactions were omitted.

			AUDPC		Phenological state		Chlorophyll		Plant height		Spike length	
	Source of variation	df	Wald %	Chi pr.	Wald %	Chi pr.	Wald %	Chi pr.	Wald %	Chi pr.	Wald %	Chi pr.
	Bobwhite vs. Frisal	1	2.6	<0.001***	83.8	<0.001***	79.8	<0.001***	79.3	<0.001***	5.7	<0.001***
Variety Bobwhite (BW)	Fungicide vs. Mildew within BW	1	35.1	<0.001***	2.0	0.004**	0.1	0.431	0.9	0.009**	0.1	0.711
	BW/GM in Fungicide	1	1.2	<0.001***	0.0	0.727	3.4	<0.001***	2.1	<0.001***	21.4	<0.001***
	Pm3b#1/2 in Fungicide	1	0.0	0.856	1.7	0.008**	4.7	<0.001***	0.1	0.470	0.0	0.890
	Natural vs. Artifical within BW	1	0.1	0.298	1.0	0.042*	1.1	<0.001***	0.1	0.377	0.7	0.226
	BW/GM in Natural	1	13.6	<0.001***	0.4	0.207	0.7	0.006**	0.4	0.083	6.4	<0.001***
	Pm3b#1/2 in Natural	1	1.6	<0.001***	3.7	<0.001***	3.4	<0.001***	1.8	<0.001***	12.2	<0.001***
	BW/GM in Artificial	1	20.4	<0.001***	0.5	0.154	0.5	0.025*	0.1	0.419	10.2	<0.001***
	Pm3b#1/2 in Artificial	1	0.0	0.576	1.0	0.039*	3.2	<0.001***	0.9	0.007	4.3	0.004**
Variety Frisal	Fungicide vs. Mildew within Frisal	1	19.8	<0.001***	0.1	0.505	0.0	0.830	1.9	<0.001***	1.0	0.164
	Frisal/GM in Fungicide	1	0.0	0.896	0.0	0.973	0.0	0.679	2.7	<0.001***	13.1	<0.001***
	A9/A13 in Fungicide	1	0.3	0.077	0.1	0.510	0.0	0.805	0.9	0.009**	1.2	0.122
	Natural vs. Artifical within Frisal	1	0.0	0.883	0.7	0.080	0.2	0.172	0.1	0.370	0.9	0.193
	Frisal/GM in Natural	1	0.1	0.293	0.1	0.531	0.1	0.355	0.2	0.188	3.7	0.007**
	A9/A13 in Natural	1	0.1	0.451	0.1	0.567	0.3	0.085	0.6	0.032*	1.5	0.088
	Frisal/GM in Artificial	1	0.0	0.544	0.1	0.492	0.0	0.660	0.0	0.954	2.7	0.022*
	A9/A13 in Artificial	1	0.0	0.723	0.0	0.691	0.0	0.497	1.3	0.001***	1.0	0.166
Diversity	GM richness	1	0.0	0.729	0.0	0.696	0.2	0.107	0.0	0.741	0.6	0.271
	GM concentration	1	0.2	0.166	0.0	0.791	0.0	0.527	0.6	0.027*	0.0	0.847
	GM concentration	1	0.1	0.444	0.2	0.646	0.2	0.119	0.3	0.155	0.2	0.542
	GM richness	1	0.1	0.228	0.0	0.915	0.1	0.452	0.4	0.085	0.5	0.349

Table S2. Summary of REML tables of seed yield, vegetative biomass, spike number, seed number and thousand seed weight (TSW). A hierarchical model was used to test for fungal infection treatment and line effects. GM richness and GM concentration were alternated. Random terms are not included in the table since their variance components are estimated directly in the REML analyses. The percentage of Wald statistic thus was calculated only for the total of the fixed effects. The total is smaller than 100% because complex interactions were omitted.

			Seed yield		Vegetative mass		Spike number		Seed number		TSW	
	Source of variation	df	Wald %	Chi pr.	Wald %	Chi pr.	Wald %	Chi pr.	Wald %	Chi pr.	Wald %	Chi pr.
	Bobwhite vs. Frisal	1	33.2	<0.001***	25.5	<0.001***	0.1	0.835	8.9	0.005**	46.5	<0.001***
Variety Bobwhite (BW)	Fungicide vs. Mildew within BW	1	0.8	0.275	0.4	0.577	0.3	0.742	1.2	0.303	10.1	<0.001***
	BW/GM in Fungicide	1	17.4	<0.001***	15.6	<0.001***	14.3	0.017*	25.1	<0.001***	0.9	0.007**
	Pm3b#1/2 in Fungicide	1	1.7	0.107	0.5	0.512	7.8	0.078	0.3	0.619	13.2	<0.001***
	Natural vs. Artifical within BW	1	1.1	0.205	4.9	0.046*	15.0	0.015*	2.0	0.183	0.0	0.734
	BW/GM in Natural	1	1.2	0.179	2.3	0.176	2.7	0.296	5.3	0.032*	0.8	0.011*
	Pm3b#1/2 in Natural	1	15.5	<0.001***	9.6	0.005**	0.0	0.932	11.6	0.002**	7.1	<0.001***
	BW/GM in Artificial	1	0.9	0.238	2.4	0.168	1.5	0.441	3.3	0.092	0.3	0.114
	Pm3b#1/2 in Artificial	1	8.0	<0.001***	6.0	0.028*	1.0	0.529	2.8	0.118	7.0	<0.001***
Variety Frisal	Fungicide vs. Mildew within Frisal	1	1.4	0.147	3.8	0.079	6.5	0.107	1.5	0.254	0.2	0.205
	Frisal/GM in Fungicide	1	2.6	0.048*	1.3	0.298	0.0	0.983	5.4	0.031*	0.3	0.152
	A9/A13 in Fungicide	1	0.7	0.323	0.1	0.735	0.0	0.961	2.4	0.144	1.3	0.001***
	Natural vs. Artifical within Frisal	1	0.4	0.435	0.2	0.718	0.2	0.759	0.2	0.668	0.2	0.163
	Frisal/GM in Natural	1	0.6	0.356	0.2	0.716	2.0	0.376	2.4	0.145	0.8	0.012
	A9/A13 in Natural	1	0.2	0.546	0.3	0.618	2.7	0.301	2.1	0.172	1.7	<0.001***
	Frisal/GM in Artificial	1	1.0	0.232	0.2	0.697	0.2	0.774	3.2	0.096	1.0	0.005**
	A9/A13 in Artificial	1	0.5	0.369	0.0	0.865	2.6	0.312	3.0	0.106	2.3	<0.001***
Diversity	GM richness	1	0.2	0.601	0.2	0.670	0.4	0.691	0.1	0.731	0.3	0.097
	GM concentration	1	0.6	0.354	0.3	0.596	0.3	0.745	0.8	0.417	0.0	0.533
	GM concentration	1	0.7	0.740	0.0	0.918	0.0	0.977	0.1	0.716	0.1	0.514
	GM richness	1	0.7	0.312	0.6	0.501	0.7	0.608	0.7	0.423	0.3	0.100

Table S3. Summary of REML tables of AUDPC, phenological state, chlorophyll content, plant height and spike length. A factorial model was used to test the fungal infection treatment and line effects. GM richness and GM concentration were alternated. Random terms are not included in the table since their variance components are estimated directly in the REML analyses. The percentage of Wald statistic thus was calculated only for the total of the fixed effects. The total is smaller than 100% because diversity contrasts (identical to Tables S1 and S2) and complex interactions were omitted.

			AUDPC		Phenological state		Chlorophyll		Plant height		Spike length	
	Source of variation	df	Wald %	Chi pr.	Wald %	Chi pr.	Wald %	Chi pr.	Wald %	Chi pr.	Wald %	Chi pr.
	Bobwhite vs. Frisal	1	2.6	<0.001***	83.8	<0.001***	79.8	<0.001***	79.2	<0.001***	5.7	<0.001***
Variety Bobwhite (BW)	Fungicide vs. Mildew within BW	1	35.1	<0.001***	2.0	0.004**	0.1	0.431	0.9	0.009**	0.1	0.711
	BW/GM in Fungicide	1	0.1	0.294	1.0	0.041*	1.1	<0.001***	0.1	0.377	0.8	0.223
	Pm3b#1/2 in Fungicide	1	28.9	<0.001***	0.4	0.177	3.7	<0.001***	1.8	<0.001***	35.6	<0.001***
	Natural vs. Artifical within BW	1	0.3	0.044*	6.0	<0.001***	11.2	<0.001***	2.2	<0.001***	10.0	<0.001***
	BW/GM in Natural	1	6.0	<0.001***	0.4	0.172	0.8	0.003**	0.6	0.026*	2.1	0.041*
	Pm3b#1/2 in Natural	1	0.2	0.087	0.0	0.784	0.1	0.326	0.5	0.041*	5.5	0.001***
	BW/GM in Artificial	1	0.3	0.047	0.0	0.898	0.0	0.684	0.1	0.512	0.2	0.524
	Pm3b#1/2 in Artificial	1	1.0	<0.001***	0.4	0.176	0.0	0.880	0.1	0.440	1.1	0.147
Variety Frisal	Fungicide vs. Mildew within Frisal	1	19.8	<0.001***	0.1	0.505	0.0	0.830	1.9	<0.001***	1.0	0.164
	Frisal/GM in Fungicide	1	0.0	0.879	0.7	0.080	0.2	0.172	0.1	0.369	0.9	0.191
	A9/A13 in Fungicide	1	0.1	0.298	0.1	0.459	0.0	0.576	1.5	<0.001***	17.2	<0.001***
	Natural vs. Artifical within Frisal	1	0.1	0.210	0.2	0.347	0.2	0.126	2.7	<0.001***	3.6	0.007**
	Frisal/GM in Natural	1	0.0	0.572	0.1	0.577	0.1	0.378	1.3	0.002**	2.3	0.035*
	A9/A13 in Natural	1	0.1	0.200	0.0	0.890	0.1	0.434	0.0	0.958	0.0	0.994
	Frisal/GM in Artificial	1	0.0	0.771	0.0	0.955	0.0	0.752	0.1	0.371	0.0	0.779
	A9/A13 in Artificial	1	0.1	0.432	0.0	0.898	0.0	0.453	0.1	0.461	0.0	0.808

Table S4. Summary of REML tables of seed yield, vegetative biomass, spike number, seed number and thousand seed weight (TSW). A factorial model was used to test the fungal infection treatment and line effects. GM richness and GM concentration were alternated. Random terms are not included in the table since their variance components are estimated directly in the REML analyses. The percentage of Wald statistic thus was calculated only for the total of the fixed effects. The total is smaller than 100% because diversity contrasts (identical to Tables S1 and S2) and complex interactions were omitted.

			Seed yield		Vegetative mass		Spike number		Seed number		TSW	
	Source of variation	df	Wald %	Chi pr.	Wald %	Chi pr.	Wald %	Chi pr.	Wald %	Chi pr.	Wald %	Chi pr.
	Bobwhite vs. Frisal	1	33.2	<0.001***	25.5	<0.001***	0.1	0.835	8.9	0.005**	46.5	<0.001***
Variety Bobwhite (BW)	Fungicide vs. Mildew within BW	1	0.8	0.275	0.4	0.577	0.3	0.742	1.2	0.303	10.1	<0.001***
	BW/GM in Fungicide	1	1.1	0.204	5.0	0.045*	15.0	0.014*	2.1	0.181	0.0	0.727
	Pm3b#1/2 in Fungicide	1	12.9	<0.001***	16.2	<0.001***	14.7	0.015*	27.6	<0.001***	0.1	0.407
	Natural vs. Artifical within BW	1	21.8	<0.001***	7.7	0.012*	0.9	0.549	6.9	0.014*	26.7	<0.001***
	BW/GM in Natural	1	6.6	0.002**	4.1	0.070	3.7	0.223	5.8	0.024*	1.9	<0.001***
	Pm3b#1/2 in Natural	1	2.9	0.039*	8.2	0.010**	7.5	0.084	6.3	0.019*	0.6	0.029*
	BW/GM in Artificial	1	0.0	0.895	0.0	0.997	0.1	0.846	0.1	0.739	0.0	0.545
	Pm3b#1/2 in Artificial	1	0.6	0.337	0.2	0.684	0.4	0.700	1.5	0.255	0.0	0.979
Variety Frisal	Fungicide vs. Mildew within Frisal	1	1.4	0.147	3.8	0.079	6.5	0.107	1.5	0.254	0.2	0.205
	Frisal/GM in Fungicide	1	0.4	0.435	0.2	0.717	0.2	0.759	0.2	0.668	0.2	0.163
	A9/A13 in Fungicide	1	3.8	0.018*	1.3	0.300	1.1	0.504	10.7	0.002**	1.9	<0.001***
	Natural vs. Artifical within Frisal	1	1.4	0.152	0.4	0.560	3.7	0.226	7.5	0.010**	5.2	<0.001***
	Frisal/GM in Natural	1	0.4	0.454	0.4	0.589	0.6	0.621	0.3	0.624	0.1	0.319
	A9/A13 in Natural	1	0.0	0.848	0.0	0.997	1.6	0.425	0.0	0.975	0.0	0.548
	Frisal/GM in Artificial	1	0.0	0.831	0.0	0.983	0.4	0.690	0.0	0.852	0.0	0.794
	A9/A13 in Artificial	1	0.0	0.842	0.1	0.814	0.0	0.980	0.0	0.872	0.0	0.693

CHAPTER 3

Mixtures of genetically modified wheat lines outperform monocultures

S.L. Zeller, O. Kalinina, B. Schmid, *Manuscript*

Fig. 4: Powdery mildew infection of a non-GM Bobwhite line in the field trial 2009

Abstract

Biodiversity research shows that diverse plant communities are more stable and productive than monocultures. Multiple pathogen resistance within a population of plants may slow down the spread of disease and reduce the risk of an individual pathogen to dominate the system. Genetically modified (GM) plants that differ only in a single resistance gene are a suitable model system to study the influence of mixtures on plant resistance to pathogens and performance.

We grew three wheat lines, two of them with a Pm3 transgene against powdery mildew, in monocultures and mixtures of two. Phenotypic measurements were taken at the level of individual plants and the population level.

We found that plant resistance to mildew increased with both GM-richness and GM-concentration in plots. Populations with two GM lines had 34.6% less mildew infection and as a consequence 7.3% higher seed yield than from the means of the two monocultures with each single GM line. We conclude that mixtures of plants that differ in their pathogen resistance should be considered. This approach seems to be more promising than stacking transgenes within a single plant where costs of resistance may accumulate. Furthermore, it conceivable that resistant pests develop more slowly in diverse multi-line systems than in single-line systems of a super-resistant plant with stacked transgenes.

Introduction

Since the mid-20eth century, the green revolution allowed agricultural yields to increase continuously, for example in bread wheat in Europe from about 1.5 t in 1950 to 7 t of grain yield per ha in 1996 and are stagnating since (Brisson et al. 2010). Fertilizer, pesticides and new crop varieties contributed to this dramatic increase in yields (Conway 1997). However, the impact of this development on the environment has also been considerable and unfortunately often negative (Tilman et al. 2001). Organic farming, on the other hand, allowed to reduce the input of agrochemicals but until now only at the cost of reduced yields (Maeder et al. 2002).

Genetic engineering may hold solutions to this problem. For example, crop plants with introduced resistance traits may help to reduce pesticide use while maintaining or even increasing yields (Borlaug 2000). Some of these genetically modified (GM) crops were so successful that they are currently planted on large areas (James 2009). This leads to a high selection pressure on the pests to overcome the resistance by evolution of new genotypes (Tabashnik et al. 2009; Powles 2010), which in turn may reduce the advantages of GM crops. Efforts are being made to slow down the evolution of such new pest genotypes. Besides refuge strategies, the combination of several GM traits within a single plant, also known as pyramiding or stacking, was promoted (Bravo and Soberon 2008). However, the sustainability of this approach is questionable because it is conceivable that "super-pests" may evolve that overcome such multiple resistance. Another problem, that has not been addressed much, is the increased defence costs that multiple resistances causes for the plant.

Here we suggest that one solution to these problems could be using mixtures of lines with different but complementary resistance traits. This at the same time allows low levels of pathogens to survive on some plants, thus reducing the selection pressure on these to evolve super-pests and avoids increased costs for multiple defences for the individual plants. The integrated defence response against multiple pathogens would thus be shifted from the individual to the population level.

Ecological theory and results of recent biodiversity experiments suggest this line of argumentation. In grassland biodiversity experiments, productivity generally increases with diversity (Tilman et al. 1996; Hector et al. 1999; Roscher et al. 2005). One of the reasons for this is reduced pathogen susceptibility of diverse systems (Zhu et al. 2000). Particular pathogens are less likely to become dominant in a diverse system because their particular hosts all occur at low abundance. Only generalist pathogens would be able to thrive in diverse systems of hosts, and such generalists usually are less efficient in overcoming the defence of a particular host due to trade-offs among the different functions needed to overcome the defences of a diverse host community.

While ecologists are currently investigating the mechanisms by which species-rich plant communities have lower pathogen abundance and higher yields (Maron et al. 2011), agronomists came across similar phenomena some time ago, albeit at the between-variety, within species level. Mixtures of several varieties of the same crop species can have higher yields than monocultures of single varieties (Browning and Frey 1969; Wolfe 1985). However, diversity strategies have rarely been used so far, probably mainly for technical reasons such as uniformity requirements for varieties and seed material and harvesting efficiency (Smithson and Lenne 1996). In part these technical difficulties may be overcome with better harvesting technology. However, plants that are phenotypically similar and differ only in a few complementary resistance traits could be produced by gene technology and thus provide an easier solution to the problem.

We experimentally compared wheat (*Triticum aestivum* L.) stands consisting of single lines with mixed stands. The lines differed only in their resistance to powdery mildew (*Blumeria graminis* f.sp. *tritici* (DC.) Speer), which was possible due to the introduction of a single gene using gene technology. One non-transgenic control line and two transgenic (GM) lines of spring wheat variety Bobwhite were used in the experiment. Mildew infection, plant production and seed yield were assessed at the level of the individual plants and the stand to test their response to increasing GM-richness (0, 1, or 2 GM lines) or GM-concentration (0, 50, 100% of individuals from GM lines) of the stands. We hypothesized that higher GM-concentration and especially higher GM-richness reduce infection with powdery mildew (H1) and therefore increase seed yield (H2) at stand level. To test the underlying mechanism, we analyzed the contributions of the different wheat lines to the population-level performance of the stand (H3) and the individual-level performance of the different wheat lines (H4).

Materials and methods

Genetically modified wheat

We used two transgenic wheat lines, derived from different transformation events of Bobwhite SH 98 26 and carrying transgenes *Pm3a* or *Pm3b*, and the control line Bobwhite SH 98 26 (von Burg et al. 2010; Peter et al. 2010; Zeller et al. 2010). *Pm3a* and *Pm3b* confer race-specific resistance to powdery mildew and were cloned from hexaploid wheat (Yahiaoui et al. 2004). The lines were generated by biolistic transformation of spring wheat variety Bobwhite SH 98 26 (Pellegrineschi et al. 2002). The plasmids pAHC17+NotI (PMI) and pAHC17+3NotI (*Pm3b*) were used as vectors (Christensen and Quail 1996; Travella et al. 2006). After NotI (for *Pm3b*) or NotI/HindIII (for PMI) digestion, only the desired fragments, but no vector sequences, were co-bombarded into wheat.

The *Pm3a* and *Pm3b* genes were cloned under the control of the Zea mays L. (maize) ubiquitin promoter (Christensen and Quail 1996) and transformants were selected on mannose-containing media using the phosphomannose isomerase (PMI)-coding gene as selectable marker (Reed et al. 2001). Absence/presence of the transgenes was confirmed by Southern hybridization analysis (Southern 2006) using probes from the PM3B (bp 1231–1956 as referred to the GenBank accession AY325736) and PMI (bp 271–810 as referred to the GenBank accession AAC74685) encoding region. The GM lines contained the Pmi gene as well as one complete copy of *Pm3b*. For transgenic lines we can not exclude the presence of fragments from the coding genes or promoter/terminator regions which were not covered by the probes used in Southern blotting. T0 transformants were regenerated, multiplied to T4 and used for the field experiments. The seeds used in this study were thus obtained from GM lines that had passed through four generations of sexual reproduction. A gene's position on the chromosome can influence its expression (Henikoff 1979; Gottschling et al. 1990). The expression level of the *Pm3a* and *Pm3b* transgenes in the two GM lines was therefore assessed by qRT-PCR using RNA isolated from leaves of seedlings grown in the glasshouse. The transgenes expression differed significantly between the two GM lines. *Pm3a* (#1) was 6–45 times and *Pm3b* (#1) 11–130 times higher expressed than in wheat line Chul which harbours this gene naturally (Brunner et al. 2011 and S. Brunner personal communications). As control for equal amount and quality of template cDNA, the expression levels of the Mlo gene (Yu et al. 2005) were determined.

Field experiment

The field experiment took place at an agricultural research station in Zurich-Reckenholz, Switzerland, from March–July 2009. Four replicate blocks, each with six 3 x 1.08 m plots, were sown with *Pm3a*, *Pm3b* and Bobwhite SH 98 26 monocultures and three 1.1 mixtures (*Pm3a*/Bobwhite, *Pm3b*/Bobwhite and *Pm3a*/*Pm3b*). In each plot, 400 seeds were sown in six rows with a distance of 17.8 cm between rows using an Oyjord plot drill system (Wintersteiger AG, Ried, Austria). The experimental plots were alternated with triticale plots in a chessboard like design to eliminate possible neighbour effects. To allow uniform infection by powdery mildew, single rows of the susceptible winter wheat variety Kanzler were planted on both sides of each plot. Powdery mildew infection occurred naturally and evenly throughout the experiment.

All seeds were treated with the fungicide Jockey (167g l^{-1} Fluquinconazole, 34 g l^{-1} Prochloraz; Omya Agro AG, Safenwil, Switzerland) before sowing. The amount of mineralized nitrogen, determined at the end of February in the top 100 cm of the soil was 35.1 and 47.6 kg N ha^{-1} in blocks 1/2 and 3/4, respectively. Nitrogen fertilizer was applied the day before sowing (40 kg N ha^{-1} in blocks 1/2, 30 kg N ha^{-1} in blocks 3/4) and again 30 kg N ha^{-1} ("Ammonsalpeter 27.5", Lonza, Visp, Switzerland) at the phenological stage 22–29 (Zadoks et al. 1974). The natural field soil provided the plants with sufficient phosphorous, potassium and magnesium (75, 182 and 213 mg kg^{-1}). All plots were sprayed with the herbicide cocktail Concert SX (40% Thifensulfurone, 4% Metusulfurone-methyl; Stähler Suisse AG, Zofingen, Switzerland) and Starane super (120 g l^{-1} Bromoxynil, 120 g l^{-1} Ioxynil, 100 g l^{-1} Fluroxypyr-metilheptil-ester; Omya Agro AG, Safenwil, Switzerland) in the beginning of May. All plots were treated twice with the insecticide Karate Zeon (100g l^{-1} Lambda-Cyhalothrin; Syngenta Agro AG, Dielsdorf, Switzerland) against the wheat stem fly (Chlorops pumilionis Bjerk.) in the beginning of May and 2 weeks later. Due to weed-infestation the whole trial was sprayed with Puma extra (69 g l^{-1} Fenoxaprop-P-ethyl, 75 g l^{-1} Mefenpyr-Diethyl; Omya Agro AG, Safenwil, Switzerland).

In each plot, ten individual plants were marked shortly after germination. These individuals were distributed evenly over the 3 m plot length and randomly among the four inner rows. This allowed us to obtain a representative sample of the entire plot population while excluding edge effects.

Response variables

We measured six phenotypic traits on individual plants (plant level) and five traits on entire plots (population level). Individual plants were assessed for the degree of powdery mildew infection (Eyal et al. 1987) 44, 59 and 78 days after germination. Based on these time points, the Area under Disease Progress Curve, AUDPC, was calculated (Jeger and Viljanen-Rollinson 2001). Furthermore, phenological stage (Zadoks et al. 1974) and height were assessed 59 and 78 days, respectively, after germination for each plant. At the end of the growing season, height was recorded again and then all individual plants were cut at ground level and separated into vegetative and reproductive parts (spikes). Vegetative and reproductive parts were dried at 80 and 25

C°, respectively, and weighed. The reproductive parts were threshed to obtain seeds and determine total seed mass per plant, here referred to as seed yield. Finally, the seed mass of the individual plants was divided by the number of seeds and multiplied by thousand to calculate the thousand seed weight (TSW).

Two non-destructive measurements were conducted at the population level. Leaf Area Index (LAI) was measured on the western side of each plot 25 and 35 days after germination (LAI 2000 Plant Canopy Analyser, LI-COR Biosciences; Lincoln, USA). It consisted of two measurements close to an inner row and one between the rows as well as a control measurement above the canopy. To assess differences in flowering time, the percentage of flowering spikes in each plot was determined 64 days after germination. At this time, all plots had flowering spikes. A subplot of 50 x 72.2 cm was harvested in the same place were the LAI was measured in each plot. These subplots were placed 50 cm from the western edge of the plot and excluded the two outer rows. The harvested material was separated into vegetative and reproductive parts to determine biomass, seed yield and thousand seed weight at population level. The latter was determined on a sample of 1,000 seeds.

Data analysis

We analysed the data of individual plants and populations separately by mixed-model analysis of variance using the REML (Restricted Maximum Likelihood) method. We used the statistical software GenStat (VSN International Ldt.). The critical significance level was 0.05 in all analyses. For the population data, where the sample size was small, we also present and discuss results significant at the 0.1 level (Peto et al. 1976; Toft and Shea 1983). The results of the mixed-model analyses were summarized in tables for all variables (see Tables S1 and S2). Residual plots were examined to identify outliers and to check if the assumptions of normality and homoscedasticity were fulfilled. Two contrasts were made to test whether diversity of GM plants (GM-richness) within each plot or the concentration of GM plants (GM-concentration) differed for the measured traits. Since these two contrasts were partly confounded with each other, their fitting sequence was alternated in two statistical models. Unless otherwise indicated, we only present results from the model where GM-richness was fitted first, i.e. ignoring the potential confounding with the term GM-concentration. Predicted means and standard errors from the REML-output were used to draw figures.

Since several of the measured traits were correlated with each other, we also performed a multivariate analysis of variance (MANOVA) to test for the overall significance of treatment effects. For the individual plant data the six traits, AUDPC, phenological state, plant height, biomass, seed mass and TSW, were combined in the MANOVA. For the population data the five traits, LAI, flowering time, biomass, seed mass and TSW, were combined.

To compare mixtures with monocultures of wheat lines, a deviation or D-value (Loreau 1998) was calculated for each plot containing a line mixture. For this, the mean of the two monocultures was first subtracted from the mixture and the resulting value then divided by the mean of the two monocultures. A D-value greater than 0 indicates, for example, that the yield of a mixture is higher than what would be expected from the mean of the monocultures. The opposite would be true for a negative D-value. We calculated D-values for powdery mildew infection, population biomass, seed yield and TSW. Original data (not predicted means) were used to calculate D-values and to draw Figures 3 and 4.

To investigate mechanisms that might explain the observed treatment effects, we added covariates to the above analyses. Powdery mildew infection had the best explanatory power for variation in the other traits and results of REML models with this covariate are thus also presented.

Results

Powdery mildew

Powdery mildew infection as measured by AUDPC at the individual plant level decreased with increasing GM-richness and GM-concentration of plots (Figure 1A; P<0.001; see Supplementary Table S1). Both contrasts were highly significant if fitted first (GM-richness: P<0.001; GM-concentration: P<0.001) or second (GM-richness: P=0.038; GM-concentration: P=0.031) in the statistical model. Plots containing two GM lines had 65.1% and plots containing on GM line had 31.7% lower mildew infection than non-transgenic control plots. Plots with 50% GM plants had 31.7% and plots with 100% GM plants had 52.8% lower mildew infection than plots without GM plants. No significant difference between the two GM lines *Pm3a* and *Pm3b* was detected (P=0.141).

All mixtures were less infected by mildew than expected from the means of the monocultures (Figure 3). D-values were –0.072, –0.144 and –0.345 for the mixtures BW/*Pm3a*, BW/*Pm3b* and *Pm3a*/*Pm3b*, respectively. This means that plants in plots with BW/*Pm3a* plants had 0.3%, plots with BW/*Pm3b* 20.7 and plots with both GM lines had 34.6% less powdery mildew than expected from the corresponding monoculture means.

Other traits measured at plant level

The phenological development of GM plants measured 59 days after germination was on average not significantly different from that of control plants (Figure 1B and Supplementary Table S1). However, *Pm3b* developed significantly faster than *Pm3a* (difference = 2.2 points on Zadoks Scale, $P<0.001$). This means that an introduced transgene can influence the phenological development of a plant.

Individual plants in Bobwhite control plots were significantly shorter than in plots harbouring GM plants (Figure 1C; difference = 3.8cm; $P=0.014$). Plant height increased with GM-richness and GM-concentration (sum of the two contrasts significant at $P=0.013$). However, the individual contrasts were only significant if fitted first in the statistical model (GM-richness: $P=0.013$; GM-concentration: $P=0.013$).

Pm3a had significantly more biomass than *Pm3b* (Figure 1D; difference= 0.55 g/plant; $P=0.036$). There was a trend towards higher biomass with increased GM-richness ($P=0.099$) but GM-concentration did not influence the biomass of individual plants. *Pm3a* had a marginally higher seed yield than *Pm3b* ($P=0.055$) and GM-richness marginally increased seed yield as well ($P=0.092$). *Pm3a* had significantly more (data not shown, $P=0.003$) but lighter seeds than *Pm3b* (Figure 1F, difference = 5.4 g TSW; $P=0.003$). TSW increased with either GM-richness or GM-concentration if the corresponding contrast was fitted first in the statistical model (GM-richness: $P=0.023$; GM-concentration: $P=0.047$) but not if it was fitted second.

The multivariate analysis for the individual plant data showed that either GM-richness or GM-concentration were significant if fitted first ($P=0.002$ for both, Table S1) but not if fitted second. Furthermore, *Pm3a* significantly differed from *Pm3b* in the multivariate analysis ($P=0.001$).

Population level data

The LAI measured at the beginning of the growing season (25 days after germination) decreased with increasing GM-concentration (Figure 2A and Supplementary Table S2; GM-concentration: $P=0.01$ if fitted first and $P=0.028$ if fitted second). However, this effect had disappeared 35 days after germination. On day 64 after germination, plots with high GM-concentration had less flowering spikes than plots with low GM-concentration (Figure 2B; $P=0.005$). Fitted after GM-concentration, GM-richness also affected the number of flowering spikes (P=0.012), however, mainly because plots with one GM line flowered later than BW monocultures. Furthermore, plots with *Pm3a* had significantly fewer flowering spikes than plots with *Pm3b* (P<0.001). This result is consistent with the individual plant data, were *Pm3a* was shown to develop more slowly than *Pm3b*.

The aboveground biomass of the populations did not differ significantly among the six lines and mixtures (Figure 2C). However, a positive D-value of 0.062 (Figure 4A) indicated that the GM-GM mixture had a higher biomass than expected from the mean of the two GM monocultures. Clearer differences were found for seed yield (Figure 2D). Populations with high GM-richness had higher yield than populations with low GM-richness (P=0.04). In numerical values populations with two GM lines had a 16.7% higher seed yield than control lines whereas populations with only one GM line only had a 5.4% higher seed yield than control lines. A positive D-value of 0.073 indicated that the GM-GM mixture performed 7.3% better than expected from the mean of the two GM monocultures (Figure 4B). Since the mixture was also producing a higher seed yield than the better GM monoculture, there was evidence for transgressive overyielding (Schmid et al. 2008).

The TSW increased significantly with GM-richness (Figure 2E, P=0.006). Seeds from plots with two GM lines were 11.9% heavier than seeds from control plots, whereas seeds from plots with only one GM line were only 5.6% heavier than seeds from control plots. This was also reflected in positive D-values for all mixtures (Figure 4C). Similar to the individual plant data, seeds from plots containing *Pm3b* were significantly heavier than seeds from plots containing *Pm3a* (P=0.016).

In the multivariate analysis with the population level data GM-concentration was significant if fitted first or second (P=0.021 and P=0.005). GM-richness, however, was only significant if fitted second, i.e. after GM-concentration (P=0.020), indicating that given the same GM-concentration, plots with two GM lines differed from plots with only one GM line. Furthermore, plots containing *Pm3a* differed significantly from plots containing *Pm3b* (P> 0.001).

Analyses with covariate mildew infection

To assess the influence of the mildew infection on other measured traits we repeated the analysis with AUDPC as covariate. On the individual plant level, plant height and TSW were affected significantly (plant height: P=0.001; TSW: P=0.002) by AUDPC. The inclusion of the covariate fully explained the effects of GM-richness and -concentration on plant height and TSW. Thus the two contrasts were no longer significant if fitted after the covariate. However, the differences between lines *Pm3a* and *Pm3b* persisted. At the population level, biomass, seed yield and TSW were significantly influenced by the covariate. Whereas the covariate did not remove the significance of the remaining effects on plot biomass, it did explain the GM-richness and -concentration effects on seed yield and TSW at population level, which both were no longer significant if fitted after the covariate. However, the differences between plots containing line *Pm3a* vs. *Pm3b* remained significant. Overall, these results suggest that the reduced mildew infection found in plots with high GM-richness or -concentration had a positive influence on plant height, seed yield and TWS.

Discussion

Mixing GM lines reduces mildew infection and increases yield

This study demonstrates that genetically modified (GM) wheat plants behave differently if grown in single-line monocultures than if grown in mixtures with other GM lines or control lines. The performance of individual plants and of entire populations generally increased with the number of GM lines (GM-richness, ranging from 0–1–2) or with the proportion of GM plants (GM-concentration, ranging from 0–50–100%) in a plot. Thus, powdery mildew resistance increased with GM-concentration, indicating that the transgene worked as expected. Furthermore, mildew resistance also increased with GM-richness. This was probably due to the fact that the two GM lines harboured transgenes that were effective against different races of powdery mildew and thus they could complement each other in mixture and provide resistance against a wider spectrum of pathogens than if the same lines were grown in single-line mixtures. This indicates that a diversity of resistance transgenes can have a beneficial effect already at the population level, avoiding the need to stack these genes in each single plant with potentially higher fitness costs. If in mixtures a certain proportion of individual plants are resistant against a specific pathogen they can reduce the spread of infection (Browning and Frey 1969; Schmid 1994). Not only mixtures of two GM lines, but also mixtures of a GM line with a control line were less infected with powdery mildew than expected from the means of the two monocultures. In this case as well, the non-resistant plants of the control line may have profited from the protection by resistant GM neighbour plants.

Besides the resistance to powdery mildew, we assessed a number of phenotypic traits correlated with performance. Individual plants grew taller and produced larger seeds in plots with increased GM-richness or -concentration. However, at the population level we recorded a lower leaf area index at the beginning of the growing season and a later flowering time in plots with high GM-concentration. This could indicate costs of resistance (Bergelson and Purrington 1996). Nevertheless, seed size and seed yield increased with GM-richness: one of the two populations with a GM/control line mixture (*Pm3b*/BW) increased its yield by 3.8% compared the mean of single monocultures. Because the seed yield of the mixture of the two GM lines was even higher than that of the better single-GM line monoculture (yield of *Pm3b*/*Pm3a* mixture was 6.5% higher than in *Pm3b*), this can be considered as one of the rare cases of transgressive overyielding (Trenbath and Harper 1974; Harper 1977; Vandermeer 1989) were two parts of a system improve their performance by interacting with each other. Using mildew infection as a covariate in the statistical analysis explained most of the differences in performance between plots with different GM-richness or -concentration, indicating that overall it was indeed the increased mildew resistance that caused the positive effects of GM-richness and -concentration on performance, in the analyses without the covariate. Thus, overall it appears that the benefits outweighed the costs of resistance under the level of mildew attack present in the field during our experiment.

Differences among GM Lines

We asked whether the introduction of different alleles of a Pm3 transgene also affected plant performance. This was indeed the case. Even though the trait directly linked to the transgene, mildew resistance, was similar in both tested lines, we found that the phenological state and the start of flowering differed strongly between the two GM lines. Individual-plant and population-level data showed consistently that *Pm3a* took longer to reach the reproductive stage than *Pm3b*. Although at population level biomass and yield did not differ, individual *Pm3a* plants had higher biomass and marginally higher seed yield than *Pm3b*. The TSW analysis revealed that *Pm3a* had generally smaller seeds than *Pm3b*. Combining this information, it can be suggested that the slower development of *Pm3a* allowed the individual plants to stay longer in the vegetative phase, develop more biomass and produce more but smaller seeds.

Since both GM lines had similar mildew resistance, it is not likely that the performance differences described above were caused directly by the powdery mildew infections. Other potential causes could have been (Cubas et al. 1999; Filipecki and Malepszy 2006) different insertion places of the transgenes in the genome or different expression levels, possibly due to different insertion places (Zeller et al. 2010).

Obviously, introduction of a fungal resistance gene can alter the phenotype of a plant in various ways.

Individual plant or population level?

The performance of crop plants is usually assessed in large plots and therefore by averaging over a whole population of plants. In particular for agronomic traits, it is the values per area of land rather than the values per plant that should be compared. However, this can lead to situations, in which important interactions among plants within a field are overlooked, which influence the overall performance of the whole population. To study such interactions, plants need to be measured and harvested individually (see Chapter 4). We assessed three wheat lines and three mixtures using both methods and can therefore also compare them, at least for those traits that are measurable at the level of the individual plant and at the population level. The results of statistical analyses in the comparable cases were very similar (Table S1 and S2), indicating that extrapolations from individual- to stand-level performance were possible in our case. Differences in phenological development and TSW among the two GM lines were found with both methods. GM-richness and GM-concentration showed similar trends for biomass, seed yield and TSW. Only the significantly increased seed yield due to increased GM-richness at the population level would not have been predicted by the results from individual plants. The explanation might lie in the density dependence of seed yield. Individual plants can and should be used for all traits like plant height, phenological development, TSW and seed set. However, for correct estimates of biomass and seed yield, the crop density or number of tillers would have to be included in the extrapolation from individual plant to population.

Generally, individual plants proved to be a useful tool to assess the performance of genetically modified wheat. This method might be labour intensive but there are also several advantages: only a few plants need to be removed from each population. This means that the experimental plots stay intact and can be used for other purposes. Furthermore, individual plants can be handled and stored much easier than bulky harvest bags.

Conclusions and applied aspects

Our study demonstrated that mixing wheat lines that differed only in their resistance to different strains of powdery mildew reduced plant susceptibility to this pathogen (H1). This led to an increased performance of these mixtures and even to transgressiv overyielding (H2). Not only mixtures of two GM lines compared to monocultures of one GM line, but also mixtures of one GM and one control line compared to monocultures on GM and control lines showed increased mildew resistance and in most cases also higher performance. One could argue therefore, that mixing closely related plant lines could increase agricultural output. Ecological research indicates that productivity increases with diversity in most cases that have been experimentally investigated (Tilman et al. 1996; Hector et al. 1999; Roscher et al. 2005). Unfortunately, this knowledge has not found its way into agricultural practice; mainly because mixtures of different varieties are difficult to harvest. Gene technology might provide us with very similar plant lines that differ only in their resistance genes. Such mixtures could therefore be harvested without change of practise. We have only assessed mixtures of two lines, either two GM lines or mixtures of one GM and one control line. According to ecological theory, mixtures of more than two lines should lead to even better results. In the future, results of such mixture experiments should be compared to lines that have several resistance genes stacked within the same plant. We hypothesize, that costs of resistance would accumulate in such plants and that they could not profit from the synergistic effects of mixtures.

Furthermore, the evolution of resistant pathogens should be studied. Some studies report that resistances may develop faster if single-gene plants that harbour different resistance genes are planted next to double-gene plants (Zhao et al. 2005). However, it is also possible that the resistance development is slower in mixtures due to the lower pathogen population size (Chin and Wolfe 1984).

The comparison of the two GM lines that harbour a different allele of the Pm3 gene revealed a number of phenotypic changes in performance-related traits which might have been of pleiotropic origin (H3). Several studies report that genetically modified plants might differ in many traits even if they share very similar transgenes (Snow et al. 2005; Filipecki and Malepszy 2006).

Finally we checked whether results obtained from individual plants can help to predict the performance of entire populations (H4). We conclude that such measurements can be very useful for performance tests — especially when information about the variation and interactions within the population are of interest. We conclude that today's agricultural systems might become both more productive and more sustainable if the advantages of biodiversity strategies such as planting line mixtures would be considered.

Acknowledgments

We thank S. Brunner and B. Keller for seed material; the national research station Agroscope Reckenholz-Tänikon ART for setting up the field experiment and Y. Kostetskyi and numerous helpers for assistance in the field. This project was supported by the Swiss National Science Foundation and is a part of the wheat-cluster.ch, a sub-unit of the national research programme NRP 59 (SNF 405940–115607).

References

Bergelson, J. & Purrington, C. B. (1996) Surveying patterns in the cost of resistance in plants. *American Naturalist,* **148,** 536–558.

Borlaug, N. E. (2000) Ending world hunger. The promise of biotechnology and the threat of antiscience zealotry. *Plant Physiology,* **124,** 487–490.

Bravo, A. & Soberon, M. (2008) How to cope with insect resistance to Bt toxins? *Trends in Biotechnology,* **26,** 573–579.

Brisson, N., Gate, P., Gouache, D., Charmet, G., Oury, F.-X. & Huard, F. (2010) Why are wheat yields stagnating in Europe? A comprehensive data analysis for France. *Field Crops Research,* **119,** 201–212.

Browning, J. A. & Frey, K. J. (1969) Multiline cultivars as a means of disease control. *Annual Review of Phytopathology,* **7,** 355–382.

Brunner, S., Hurni, S., Herren, G., Kalinina, O., von Burg, S., Zeller, S., Schmid, B., Winzeler, M. & Keller, B. (2011) Transgenic Pm3b wheat lines show resistance to powdery mildew in the field. *Plant Biotechnology Journal,* **in press.**

Chin, K. M. & Wolfe, M. S. (1984) Selection on *Erysiphe graminis* in pure and mixed stands of barley. *Plant Pathology,* **33,** 53–545.

Christensen, A. H. & Quail, P. H. (1996) Ubiquitin promoter-based vectors for high-level expression of selectable and/or screenable marker genes in monocotyledonous plants. *Transgenic Research,* **5,** 213–218.

Conway, G. R. (1997) *The doubly green revolution: food for all in the 21st century.* Penguin Books, London/Cornell University Press, Ithaca NY.

Cubas, P., Vincent, C. & Coen, E. (1999) An epigenetic mutation responsible for natural variation in floral symmetry. *Nature,* **401,** 157–161.

Eyal, Z., Scharen, A., Prescott, J. & van Ginkel, M. (1987) *The septoria diseases of wheat: concepts and methods of disease management.* International Maize and Wheat Improvement Center, D.F. Mexico.

Filipecki, M. & Malepszy, S. (2006) Unintended consequences of plant transformation: a molecular insight. *Journal of Applied Genetics,* **47,** 277–286.

Gottschling, D. E., Aparicio, O. M., Billington, B. L. & Zakian, V. A. (1990) Position effect at *Saccharomyces cerevisiae* telomeres - reversible repression of Pol-Ii transcription. *Cell,* **63,** 751–762.

Harper, J. L. (1977) *Population biology of plants.* Academic Press, London.

Hector, A., Schmid, B., Beierkuhnlein, C., Caldeira, M. C., Diemer, M., Dimitrakopoulos, P. G., Finn, J. A., Freitas, H., Giller, P. S., Good, J., Harris, R., Hogberg, P., Huss-Danell, K., Joshi, J., Jumpponen, A., Korner, C., Leadley, P. W., Loreau, M., Minns, A., Mulder, C. P. H., O'Donovan, G., Otway, S. J., Pereira, J. S., Prinz, A., Read, D. J., Scherer-Lorenzen, M., Schulze, E. D., Siamantziouras, A. S. D., Spehn, E. M., Terry, A. C., Troumbis, A. Y., Woodward, F. I., Yachi, S. & Lawton, J. H. (1999) Plant diversity and productivity experiments in European grasslands. *Science,* **286,** 1123–1127.

Henikoff, S. (1979) Position effects and variegation enhancers in an autosomal region of *Drosophila melanogaster. Genetics,* **93,** 105–115.

James, C. (2009) *Global status of commercialized biotech/gm crops: 2009.* ISAAA, Ithaca, NY.

Jeger, M. J. & Viljanen-Rollinson, S. L. H. (2001) The use of the area under the disease-progress curve (AUDPC) to assess quantitative disease resistance in crop cultivars. *Theoretical and Applied Genetics,* **102,** 32–40.

Loreau, M. (1998) Separating sampling and other effects in biodiversity experiments. *Oikos,* **82,** 600–602.

Maeder, P., Fliessbach, A., Dubois, D., Gunst, L., Fried, P. & Niggli, U. (2002) Soil fertility and biodiversity in organic farming. *Science,* **296,** 1694–1697.

Maron, J. L., Marler, M., Klironomos, J. N. & Cleveland, C. C. (2011) Soil fungal pathogens and the relationship between plant diversity and productivity. *Ecology Letters,* **14,** 36–41.

Pellegrineschi, A., Noguera, L. M., Skovmand, B., Brito, R. M., Velazquez, L., Salgado, M. M., Hernandez, R., Warburton, M. & Hoisington, D. (2002) Identification of highly transformable wheat genotypes for mass production of fertile transgenic plants. *Genome,* **45,** 421–430.

Peto, R., Pike, M. C., Armitage, P., Breslow, N. E., Cox, D. R., Howard, S. V., Mantel, N., Mcpherson, K., Peto, J. & Smith, P. G. (1976) Design and analysis of randomized clinical-trials requiring prolonged observation of each patient .1. Introduction and design. *British Journal of Cancer,* **34,** 585–612.

Powles, S. B. (2010) Gene amplification delivers glyphosate-resistant weed evolution. *Proceedings of the National Academy of Sciences of the United States of America,* **107,** 955–956.

Reed, J., Privalle, L., Powell, M. L., Meghji, M., Dawson, J., Dunder, E., Suttie, J., Wenck, A., Launis, K., Kramer, C., Chang, Y. F., Hansen, G. & Wright, M. (2001) Phosphomannose isomerase: An efficient selectable marker for plant transformation. *In Vitro Cellular & Developmental Biology-Plant,* **37,** 127–132.

Roscher, C., Temperton, V. M., Scherer-Lorenzen, M., Schmitz, M., Schumacher, J., Schmid, B., Buchmann, N., Weisser, W. W. & Schulze, E. D. (2005) Overyielding in experimental grassland communities - irrespective of species pool or spatial scale. *Ecology Letters,* **8,** 576–577.

Schmid, B. (1994) Effects of genetic diversity in experimental stands of *Solidago altissima* - evidence for the potential role of pathogens as selective agents in plant-populations. *Journal of Ecology,* **82,** 165–175.

Schmid, B., Hector, A., Saha, P. & Loreau, M. (2008) Biodiversity effects and transgressive overyielding. *Journal of Plant Ecology,* **1,** 95–102.

Smithson, J. B. & Lenne, J. M. (1996) Varietal mixtures: A viable strategy for sustainable productivity in subsistence agriculture. *Annals of Applied Biology,* **128,** 127–158.

Snow, A. A., Andow, D. A., Gepts, P., Hallerman, E. M., Power, A., Tiedje, J. M. & Wolfenbarger, L. L. (2005) Genetically engineered organisms and the environment: Current status and recommendations. *Ecological Applications,* **15,** 377–404.

Southern, E. (2006) Southern blotting. *Nature Protocols,* **1,** 518–525.

Tabashnik, B. E., Van Rensburg, J. B. J. & Carriere, Y. (2009) Field-evolved insect resistance to Bt crops: definition, theory, and data. *Journal of Economic Entomology,* **102,** 2011–2025.

Tilman, D., Fargione, J., Wolff, B., D'Antonio, C., Dobson, A., Howarth, R., Schindler, D., Schlesinger, W. H., Simberloff, D. & Swackhamer, D. (2001) Forecasting agriculturally driven global environmental change. *Science,* **292,** 281–284.

Tilman, D., Wedin, D. & Knops, J. (1996) Productivity and sustainability influenced by biodiversity in grassland ecosystems. *Nature,* **379,** 718–720.

Toft, C. A. & Shea, P. J. (1983) Detecting community-wide patterns - estimating power strengthens statistical-inference. *American Naturalist,* **122,** 618–625.

Travella, S., Klimm, T. E. & Keller, B. (2006) RNA interference-based gene silencing as an efficient tool for functional genomics in hexaploid bread wheat. *Plant Physiology,* **142,** 6–20.

Trenbath, B. R. & Harper, J. L. (1974) Neighbor effects in genus *avena.* 2. Comparison of weed species. *Journal of Applied Ecology,* **11,** 111–125.

Vandermeer, J. (1989) *The ecology of intercropping.* Cambridge University Press, Cambridge.

Wolfe, M. S. (1985) The current status and prospects of multiline cultivars and variety mixtures for disease resistance. *Annual Review of Phytopathology,* **23,** 251–273.

Yahiaoui, N., Srichumpa, P., Dudler, R. & Keller, B. (2004) Genome analysis at different ploidy levels allows cloning of the powdery mildew resistance gene *Pm3b* from hexaploid wheat. *Plant Journal,* **37,** 528–538.

Yu, L., Niu, J. S., Chen, P. D., Ma, Z. Q. & Liu, D. J. (2005) Cloning, physical mapping and expression analysis of a wheat mlo-like. *Journal of Integrative Plant Biology,* **47,** 214–222.

Zadoks, J. C., Chang, T. T. & Konzak, C. F. (1974) Decimal Code for Growth Stages of Cereals. *Weed Research,* **14,** 415–421.

Zeller, S. L., Kalinina, O., Brunner, S., Keller, B. & Schmid, B. (2010) Transgene × environment interactions in genetically modified wheat. *PloS ONE,* **In press.**

Zhao, J. Z., Cao, J., Collins, H. L., Bates, S. L., Roush, R. T., Earle, E. D. & Shelton, A. M. (2005) Concurrent use of transgenic plants expressing a single and two *Bacillus*

thuringiensis genes speeds insect adaptation to pyramided plants. *Proceedings of the National Academy of Sciences of the United States of America,* **102,** 8426–8430.
Zhu, Y., Chen, H., Fan, J., Wang, Y., Li, Y., Chen, J., Fan, J., Yang, S., Hu, L., Leung, H., Mew, T. W., Teng, P. S., Wang, Z. & Mundt, C. C. (2000) Genetic diversity and disease control in rice. *Nature,* **406,** 718–722.
Figure 4: Photograph taken by Simon Zeller

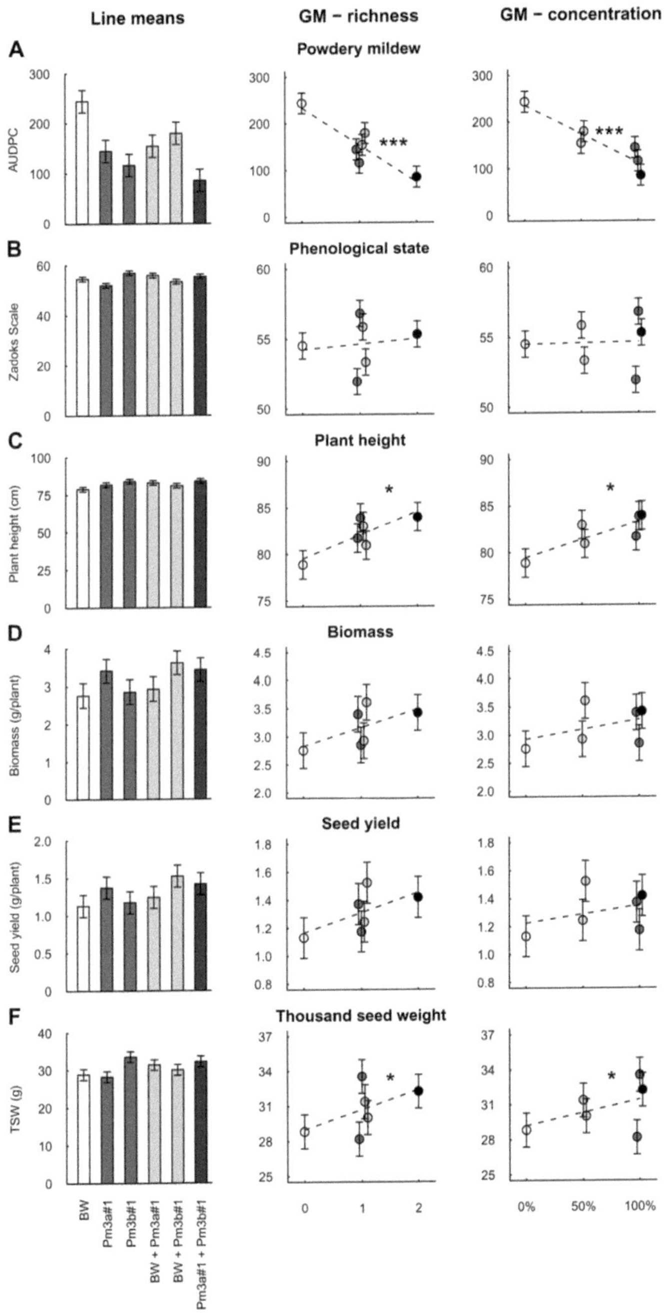
Line means
GM – richness
GM – concentration
A
Powdery mildew
AUDPC
***
***
B
Phenological state
Zadoks Scale
C
Plant height
Plant height (cm)
*
*
D
Biomass
Biomass (g/plant)
E
Seed yield
Seed yield (g/plant)
F
Thousand seed weight
TSW (g)
*
*
BW
Pm3a#1
Pm3b#1
BW + Pm3a#1
BW + Pm3b#1
Pm3a#1 + Pm3b#1
0
1
2
0%
50%
100%

Figure 1. Effects of GM-richness and GM-concentration on individual wheat plants. Line means were predicted using REML models. GM-richness consisted of the levels "no GM" "one GM" and "two GM" lines and GM-concentration of 0, 50 and 100% GM plants in a particular plot. A–F are different traits that were measured on individual plants. White bars/points=Bobwhite control; light gray bars/points=mixtures of one GM and one Control line; medium gray=GM monocultures; dark gray=mixture of two GM lines. Error bars represent ± standard error. Asterisk indicate the level of significance for the GM-richness or GM-concentration contrast (*$P<0.05$; ***$P<0.001$).

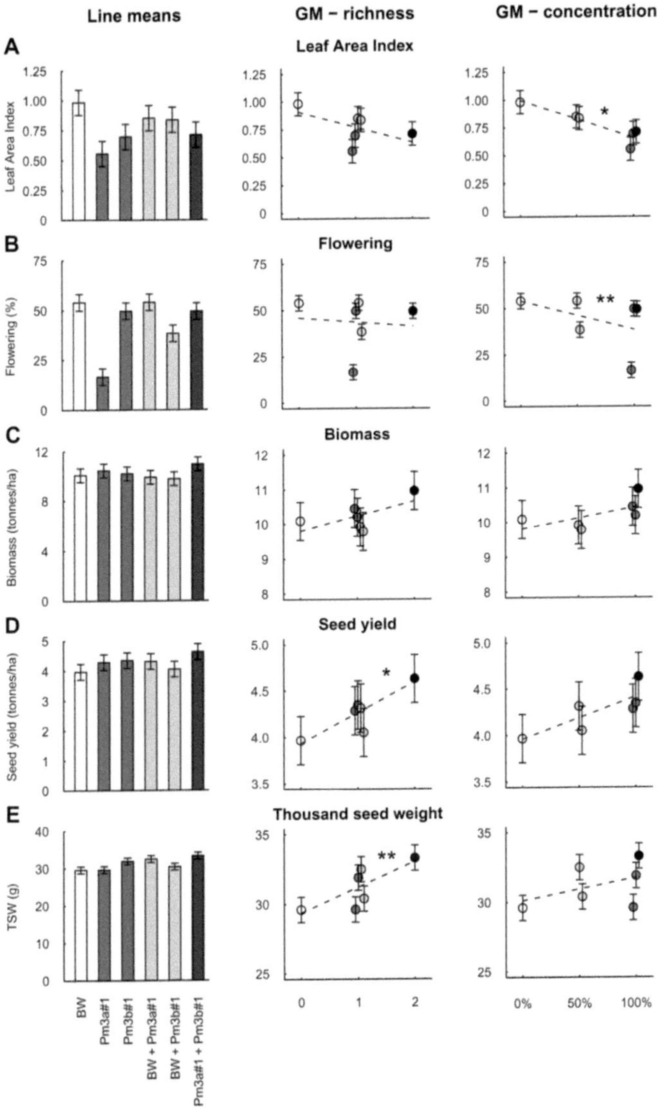

Line means
GM – richness
GM – concentration
A
Leaf Area Index
Leaf Area Index
B
Flowering
Flowering (%)
C
Biomass
Biomass (tonnes/ha)
D
Seed yield
Seed yield (tonnes/ha)
E
Thousand seed weight
TSW (g)
BW
Pm3a#1
Pm3b#1
BW + Pm3a#1
BW + Pm3b#1
Pm3a#1 + Pm3b#1
0
1
2
0%
50%
100%

Figure 2. Effects of GM-richness and GM-concentration on wheat population. Line means were predicted using REML models. GM-richness consisted of the levels "no GM" "one GM" and "two GM" lines and GM-concentration of 0, 50 and 100% GM plants in a particular plot. A–F are different traits that were measured on population level. White bars/points=Bobwhite control; light gray bars/points=mixtures of one GM and one Control line; medium gray=GM monocultures; dark gray=mixture of two GM lines. Error bars represent ± standard error. Asterisk indicate the level of significance for the GM-richness or GM-concentration contrast (*P<0.05; **P<0.01).

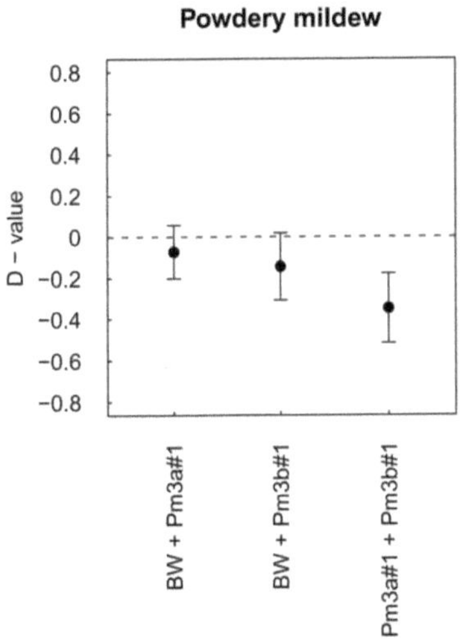

Figure 3. Deviation values for powdery mildew in different mixtures. D-values were calculated by subtracting the monoculture means from the mixture values and dividing the results by the monoculture means.

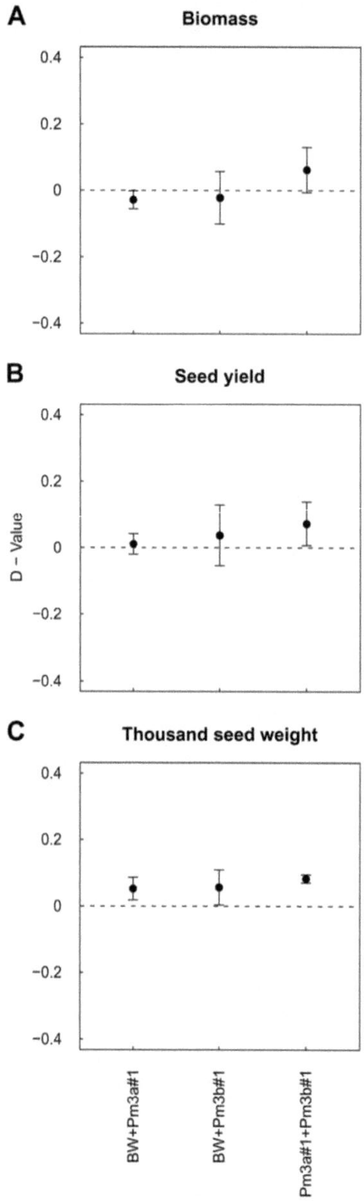

Figure 4. Deviation values for Biomass, Seed yield and Thousand seed weight (TSW) in different mixtures. D- values were calculated by subtracting the monoculture means from the mixture values and dividing the results by the monoculture means.

Table S1. Summary of Manova and REML tables of the traits measured on individual plants. Three different models were used. In model 1 all treatments are pooled. In models 2 and 3 we used different sequences for the terms GM-richness and GM-concentration: in model 2 (3) GM-richness is tested "ignoring" ("eliminating") GM-concentration and GM-concentration is tested eliminating (ignoring) GM-richness. The word "eliminating" here is equivalent to "given" or "holding constant" and indicates that the influence of the first term has already been explained and therefore no longer contributes to the explanatory power of the second term. Random terms are not included in the table since their variance components are estimated directly in the REML analyses. The percentage Rao's F-statistics or Wald statistic thus was calculated only for the total of the fixed effects.

			Manova		AUDPC		Phenological state		Plant Height		Biomass		Seed Yield		Thousand seed weight	
Model	Source of variation	df	% RaoF	F pr.	% Wald	F pr.	% Wald	F pr.	% Wald	F pr.	% Wald	F pr.	% Wald	F pr.	% Wald	F pr.
1	Treatment	5	100	<0.001	100	<0.001	100	0.006	100	0.087	100	0.176	100	0.176	100	0.009
2	GM-richness	1	36.4	0.002	83.9	<0.001	2.0	0.512	67.1	0.013	34.6	0.099	36.3	0.092	27.7	0.022
2	GM-concentration	1	6.7	0.277	11.2	0.031	1.2	0.756	8.5	0.328	0.7	0.814	3.6	0.581	0.4	0.770
3	GM-concentration	1	37.2	0.002	84.9	<0.001	0.2	0.754	67.2	0.013	16.3	0.245	12.2	0.314	19.9	0.047
3	GM-richness	1	5.9	0.341	10.3	0.038	3.0	0.513	8.4	0.330	18.8	0.215	27.8	0.136	8.1	0.187
2+3	GM-rich. x GM-conc.	1	2.3	0.778	0.0	0.902	0.7	0.478	1.4	0.690	4.9	0.515	10.2	0.356	0.1	0.876
2+3	*Pm3a* x *Pm2b*	1	44.7	0.001	4.8	0.141	87.2	0.001	23.0	0.118	59.2	0.036	48.5	0.055	52.2	0.003
2+3	GM-conc. x *Pm3a*	1	9.9	0.131	0.0	0.931	8.9	0.1	0.0	0.955	0.7	0.804	1.5	0.724	18.7	0.053
									1 value excluded							

Table S2. Summary of Manova and REML tables of the traits measured on population level. Three different models were used. In model 1 all treatments are pooled. In models 2 and 3 we used different sequences for the terms GM-richness and GM-concentration: in model 2 (3) GM-richness is tested "ignoring" ("eliminating") GM-concentration and GM-concentration is tested eliminating (ignoring) GM-richness. The word "eliminating" here is equivalent to "given" or "holding constant" and indicates that the influence of the first term has already been explained and therefore no longer contributes to the explanatory power of the second term. Random terms are not included in the table since their variance components are estimated directly in the REML analyses. The percentage of Rao's F-statistics or Wald statistic thus was calculated only for the total of the fixed effects.

			Manova		LAI		Flowering		Biomass		Seed Yield		Thousand seed weight	
Model	Source of variation	df	% RaoF	F pr.	% Wald	F pr.	% Wald	F pr.	% Wald	F pr.	% Wald	F pr.	% Wald	F pr.
1	Treatment	5	100	0.002	100	0.021	100	<0.001	100	0.516	100	0.327	100	0.021
2	GM - richness	1	9.6	0.002	33.0	0.085	1.1	0.466	43.2	0.190	79.4	0.040	54.6	0.006
2	GM - concentration	1	28.3	0.277	57.3	0.028	31.9	<0.001	2.3	0.736	3.2	0.685	5.5	0.340
3	GM - concentration	1	18.9	0.002	85.4	0.010	19.3	0.005	36.4	0.220	63.5	0.064	18.7	0.081
3	GM - richness	1	18.9	0.341	4.9	0.488	14.2	0.012	9.1	0.556	19.0	0.289	42.2	0.013
2+3	GM-rich. X GM-conc.	1	5.4	0.778	0.0	0.972	3.0	0.216	50.0	0.157	4.8	0.586	0.0	0.987
2+3	*Pm3a* X *Pm2b*	1	51.6	0.001	5.8	0.470	56.8	<0.001	0.0	0.884	9.5	0.451	39.9	0.016
2+3	GM-conc. x *Pm3a*	1	5.2	0.131	3.9	0.559	7.3	0.060	4.5	0.679	3.2	0.639	0.0	0.915

ang transformed

CHAPTER 4

Competitive performance of pathogen-resistant transgenic wheat: a phytometer approach

O. Kalinina, S.L. Zeller, B. Schmid, *Manuscript*

Fig. 5: Plot with 30 phyometers plants (orange labels) in the middle of Mai 2008

Abstract

Genetically modified c plants offer an ideal model system to study the influence of single genes on the ecological behaviour of plants. They allow, in particular, assessing costs associated with specific genes for pathogen resistance.

We used phytometers to study the competitive interactions between lines of spring wheat *Triticum aestivum* carrying transgenes for pathogen resistance. We hypothesized reduced competitive performance of GM lines due to costs of resistance under pathogen levels typically encountered in the field. The transgenes were the *Pm3b* gene from *T. aestivum* i or *Chitinase* (*Chi*) and *Glucanase* (*Glu*) genes from *Hordeum vulgare* (resistance against fungi in general). Phytometers of 15 transgenic and non-transgenic wheat lines and varieties were transplanted as seedlings into plots containing the same 15 lines and varieties as competitive environments and subject to two different soil nutrient levels.

The *Pm3b* transgene reduced the incidence of *B. graminis* compared with control lines. This difference in pathogen incidence increased with nutrient addition. *Chi* and *Chi/Glu* lines showed the same high resistance to mildew as their control.

Pm3b lines had lower competitive performance than control lines. This resulted in reduced yield and seed set, indicating that costs of carrying a single resistance gene were greater than the benefits. *Pm3b* line with the highest resistance to mildew was the weakest competitor. The line carrying the *Chi* gene did not differ from the control, but the line carrying both genes (*Chi/Glu*) also showed reduced competitive performance.

Our results suggest that single transgenes conferring pathogen resistance to plants can have large costs even in the presence of the pathogen. The reason for this could be enhanced gene expression levels in the transgenes. We suggest that altered regulation of single genes in plants can have much larger effects than typically observed in the wild, where a number of co-varying factors such as other genes may mask these single-gene effects.

Introduction

Advances in biotechnology allow the introduction of single genes against fungal pathogens into plants (Gurr and Rushton 2005; Melchers and Stuiver 2000). The resulting transgenic plants offer a perfect model system for ecologists to study the effects of single genes of resistance on the whole plant phenotype and open up new horizons for gene × environment interaction studies (Strauss et al. 2002).

It is known that resistance to a pathogen might reduce plant performance. This is usually found in the absence of pathogen attack and is often associated with allocation costs (Herms and Mattson 1992; Bergelson and Purrington 1996; Heil and Baldwin 2002). Another type of costs of resistance, which have been addressed less often, are ecological costs which arise when the interactions between a plant and its biotic or abiotic environment change in a way that affects plants negatively (Tollrian and Harvell 1999; Heil 2002; Heil and Baldwin 2002). Ecological costs are more difficult to study than allocation costs because they might not be apparent under stable growing conditions indoors or on isolated plants where the range of plant × environment interactions is limited (Heil 2002). The few studies which reported ecological costs did not control for a common genetic background of resistant and non-resistant plants and, moreover, typically used chemicals to induce resistance and thus might have been biased by side-effects of these treatments (Baldwin 1988; Heil et al. 2000; van Dam and Baldwin 2001).

To reveal effects of single resistance genes on plant performance and on gene × environment interactions, fast growing GM cereals represent a convenient model system. Transgenic plants allow controlling the genetic background to ensure that the plants differ only in one gene (Strauss et al. 2002). Furthermore, promoters used with transgenes are able to enhance gene expression hundredfold and more (Rooke et al. 2000), thus providing a possibility to consider not only the effects of gene presence but also of gene regulation on pathogen resistance and its potential costs.

The hope of introducing a single resistance gene into a plant is that the benefits of resistance may outweigh the potential costs in the presence of the pathogen. However, in this case the introduced trait may increase the relative fitness of the plants carrying the transgene compared with those lacking it, allowing the GM to become invasive in natural habitats (Tiedje et al. 1989; Ammann et al. 2000). Because the potential advantage will be a relative one and dependent on the presence of the pathogen in the environment, the competitiveness of the GM plant must be assessed against appropriate control plants across a range of environments (Crawley 1992; Fredshavn and Poulsen 1996). This has rarely been done in disease-resistant transgenic plants (Fuchs et al. 2004; Laughlin et al. 2009; Bartsch et al. 1996). Furthermore, due to the complexity of broad-range competition experiments, most studies have so far only tested a limited number of competitive interactions.

Here we used phytometers (Clements and Goldsmith 1924; Violle et al. 2009) to assess competitive interactions among a range of transgenic and non-transgenic lines and varieties of *T. aestivum* under two fertilizer levels in a full mechanistic diallel (McGilchrist 1965; van Kleunen and Schmid 2003). Phytometers are individual plants which are planted into a range of environments. Originally this approach was used to measure the quality of different test environments (Clements and Goldsmith 1924). However, it can also be used to compare the response of the different species in question to different environments (Mwangi et al. 2007) and allows one to measure a wide range of plant characteristics in a large number of environments keeping the size of the experiment reasonably small. Whereas phytometers have repeatedly been used in ecology, this is, to our knowledge, the first time they are used to study competitive behaviour and gene × environment interactions in GM plants.

We used the spring wheat *Triticum aestivum* L. variety Bobwhite SH 98 26, hence abbreviated Bobwhite, transformed with the wheat *Pm3b* gene of resistance to powdery mildew *Blumeria graminis* f.sp. *tritici* (DC.) Speer (Yahiaoui et al. 2004), and variety Frisal with introduced fungal resistance *Chi* and *Glu* genes from *Hordeum vulgare* L. (Leah et al. 1991) as a model system to study the effects of single genes of pathogen resistance on the competitive ability of plants and to establish the phytometer technique for the assessment of the ecological behaviour of a range of plant lines across a range of test environments. Specifically, we asked the following questions: (1) Do the introduced transgenes improve resistance to mildew and phytometer performance (main effects of transgenes)? (2) How do the nutrient and the competitive environments affect resistance to mildew and phytometer performance (main effects of environments)? (3) Do differences between transgenic and control lines vary across nutrient and competitive environments (overall transgene × environment interactions)? (4) Do transgenic and control lines behave differently if planted into their own rather than into different lines as competitive environments (home vs. away contrast of transgene × environment interactions)?

Materials and methods

Plant material

We used six transgenic lines derived from two maternal varieties of *T. aestivum*: Mexican variety Bobwhite and Swiss variety Frisal. Four transgenic lines (*Pm3b*#1–4) were produced by biolistic transformation of Bobwhite in different transformation events. *Pm3b*#1–3 lines carried a single copy of the transgene *Pm3b* and *Pm3b*#4 line carried one full-length and one truncated copy (Zeller et al. 2010). Their respective non-transgenic sister lines Sb#1–4 (null-segregants) were used as a control to ensure that any somaclonal variations acquired during tissue culturing were shared between transgenic and control lines. The *Pm3b* gene confers race-specific resistance to *B. graminis* and was cloned from hexaploid wheat (Yahiaoui et al. 2004). The seeds used in this study were obtained from homozygous GM and control lines that had passed through five generations of sexual reproduction. The ubiquitin promoter from *Zea mays* L. ensured that the transgene was expressed at a high level.

The GM lines derived from the variety Frisal carried a barley seed *Chitinase* or/and *β-1,3-glucanase* genes (Leah et al. 1991). One transgenic line was positively selected for chitinase expression (line A9 *Chi*) and one for chitinase and glucanase expression (line A13 *Chi/Glu*; Bieri et al. 2003). Chitinases and glucanases are known for their anti-fungal effect. The expression of these pathogenesis-related genes should result in increased quantitative resistance to *B. graminis* (Zhu et al. 1994). The seeds used for the field experiment were obtained from the sixth generation of transgenic lines A13 *Chi/Glu* and A9 *Chi*. Wheat variety Frisal was used as a control.

In addition to the 11 lines or varieties already mentioned, four further entries were used as phytometer plants: variety Bobwhite (plants that had not passed through tissue culture) and three commercialized conventional varieties of *T. aestivum*: Casana, Fiorina and Toronit (in the following "lines and varieties" are called simply "lines"). These "out-groups" were used as reference to compare differences caused by the transgenes *within* varieties with differences *between* varieties.

Field experiment

The field experiment took place in 2008 at a research station in Zurich-Reckenholz, Switzerland. The 15 lines of *T. aestivum* (six transgenic lines *Pm3b*#1–4, A9 *Chi* and A13 *Chi/Glu*, four sister lines Sb#1–4, five conventional varieties Frisal, Bobwhite, Casana, Fiorina and Toronit) were planted as test environments in a randomized complete block design with four replicate blocks. The resulting 60 plots of 7×1.08 m were split into subplots. The two edge subplots of 1×1.08 m in each plot were used for a split-plot treatment, i.e. fertilizer application vs. control. Fertilizer was applied twice, when the plants had reached phenological stage 11 (Zadoks et al. 1974) and again when they had reached stage 39, to one of the two subplots in each plot (two times 3 g N m^{-2} as "Ammonsalpeter 27.5", Lonza, Visp, Switzerland). The natural field soil provided the plants with phosphorous, potassium and magnesium (80, 235 and 234 mg kg^{-1}, respectively).

In each subplot, 400 wheat seeds were sown in six rows with a distance of 18 cm between the rows using an Oyjord plot drill system (Wintersteiger AG, Ried, Austria). All plots were sprayed with the herbicide cocktail Concert SX (40% Thifensulfurone, 4% Metusulfurone-methyl; Stähler Suisse AG, Zofingen, Switzerland) and Starane super (120 g l^{-1} Bromoxynil, 120 g l^{-1} Ioxynil, 100 g l^{-1} Fluroxypyrmetilheptilester; Omya Agro AG, Safenwil, Switzerland) at the beginning of May. *Blumeria graminis* infection occurred naturally. Each plot represented one of 15 competitive environments for the phytometers.

Phytometers

In February 2008, 3600 individual seeds of the 15 lines of *T. aestivum* (the same lines as used in the field plots) were germinated in a climate-controlled glasshouse (day/night temperature: 21/16 C°; additional light: 14 h/10 h day/night period, daily watering by hand) at the Institute of Evolutionary Biology and Environmental Studies, University of Zurich, Switzerland. In March 2008, the seedlings at the phenological stage 12 (Zadoks et al. 1974) were transplanted from the glasshouse to the field and inserted into the test environments described in the previous section. These seedlings, grown under standard conditions in the glasshouse, were now used as phytometers to assess their phenotypic response to competitive environment and fertilizer application. Since we intended to measure fitness-related traits, checking for the effect of transplanting was necessary. In our experiment, the phytometers did not differ among the wheat lines at the stage of transplanting and during early stages of growth in the field (Zadoks stage 14−15). This indicates that, if the transplanting influenced the growth of seedlings, all the phytometer lines responded to it similarly.

Thirty phytometer seedlings representing 15 lines of *T. aestivum* were introduced into each 1×1.08 m subplot. Before the phytometers were planted, already established seedlings of the competitive environment were removed from the rows to free space for five phytometers per row (six rows per subplot; Figure S1 in

Supplementary Information). The distance between neighbouring phytometer plants was 20 cm. As a result, phytometers of each of the 15 lines occurred in each of the 15 lines as competitive environments. This corresponds to a full mechanistic diallel design (McGilchrist 1965; van Kleunen and Schmid 2003). Each phytometer line was represented twice in each subplot.

Measurements

We recorded plant height and phenological stage (Zadoks et al. 1974) of all phytometer plants 53 days after planting. The incidence of powdery mildew infection was assessed 80 days after planting when infection reached its maximum. It was measured for each line of *T. aestivum* as a percentage of the plants infected with the pathogen out of all the plants. After ripening, all phytometer plants were cut at ground level and separated into vegetative and reproductive parts (spikes). All plant material was dried at 80 C° (vegetative parts) and 25 C° (reproductive parts) and weighed. We counted the spike number per phytometer plant, threshed the reproductive parts, determined the seed set per spike and obtained the total mass of seeds per plant. In the following, total mass of seeds is called yield. The phytometer data were used to characterize the competitiveness of different wheat lines. For each fertilizer treatment, we calculated the relative performance of each phytometer line by dividing its subplot mean through the mean value this phytometer line reached in its own environment (Allard and Adams 1969; McGilchrist 1965; McGraw 1985). This was used as a test for home vs. away effects, corresponding to a main-diagonal contrast within the transgene × environment interaction term (Joshi et al. 2001).

Data analysis

Data were analysed with classical mixed-model analysis of variance (ANOVA) using the statistical software GenStat (VSN International Ldt. 2010). The treatment model consisted of the factorially crossed phytometer lines and competitive environments (mechanistic diallel) and fertilizer application. The error model consisted of phytometer plants nested within subplots, subplots nested within plots and plots nested within blocks. The terms of the treatment model were tested against the appropriate terms of the error model: competitive environment varied among plots, fertilizer application among subplots and phytometer line within subplots (Figure S2 in Supplementary Information). Residual plots were examined to identify outliers and to check if the assumptions of normality and homoscedasticity were fulfilled.

The analyses were repeated with the restricted maximum likelihood (REML) approach to mixed models which yielded nearly identical results. However, because error terms sometimes were constrained to zero variance components in the REML analysis, the treatments were not tested against the appropriate error terms in these cases, resulting in less conservative significance tests.

First we analysed the originally measured variables to identify the differences in phytometer performance and the effects of competitive environment and fertilizer application. Then we analysed the relative performance values (see previous section) to compare the competitive ability of phytometer lines independently of their different performance in "pure stands" (i.e. phytometer plant in its own competitive environment). All data, original and relative measures, were log-transformed to fulfil ANOVA assumptions (normality and homoscedasticity). The binary mildew incidence data were analysed using multiple logistic regression with mixed-model analysis of deviance (Mccullagh and Nelder 1989).

In June 2008, 1093 out of 3600 phytometer plants were damaged by vandalism. These plants were excluded from the analysis of the traits measured after the damage happened. ANOVA analysis showed that the damage by vandalism occurred randomly across our phytometer plants and did not interfere with the effects of the factors of interest.

Results

Mildew incidence: Main effects of transgenes (phytometer line)

Powdery mildew incidence reached its maximum in the field 80 days after transplanting of the phytometers. Phytometers carrying the *Pm3b* transgene showed the desired decrease in mildew incidence (up to five-fold compared with control lines; $P<0.001$). The difference between *Pm3b* lines and Sb control lines explained 12.4% of the total variation in mildew incidence and exceeded the variation among the conventional varieties of *T. aestivum* (Table S1 in Supplementary Information).

The four *Pm3b* lines differed significantly from one another in mildew incidence ($P=0.01$). Line *Pm3b*#2 had the lowest and line *Pm3b*#3 had the highest scores with, respectively, 6% and 14% of phytometers infected. The four control Sb lines only marginally differed from each other in mildew incidence ($P=0.059$) and were highly susceptible to the pathogen (up to 62% of the plants infected).

The lines A9 *Chi* and A13 *Chi/Glu* had low mildew incidence, however, this was also the case for the control Frisal variety. Mildew incidence in Frisal plants never exceeded 7%. The lines derived from the Mexican variety Bobwhite had higher mildew incidence than those from Frisal variety (P<0.001).

Main effects of environments (soil nutrients and wheat environments)

Application of fertilizer increased mildew incidence in both transgenic and conventional lines of *T. aestivum* (P<0.001). The biotic environment affected mildew incidence (P=0.001) and explained 5.2% of variation in mildew incidence among the phytometer plants. Higher mildew incidence was observed for the phytometers introduced into mildew-susceptible wheat environments, Sb lines representing the most "infective" environments (P<0.001).

Overall transgene × environment interactions

The differences in mildew incidence between *Pm3b* and Sb lines increased with nutrient addition (P<0.001), mainly due to elevated mildew incidence in sister lines. The difference between *Pm3b* and Sb lines became significantly stronger in mildew-susceptible and weaker in mildew-resistant biotic environments (P=0.024).

Phytometer Performance: Main effects of transgenes (phytometer lines)

The *Pm3b* lines developed significantly lower vegetative mass, yield and seed set and had slightly advanced phenological stage and plant height compared with corresponding control Sb lines (Figure 1a), however, they did not differ from these in spike number (P=0.001 for vegetative mass, P<0.001 for yield, seed set, plant height and phenological stage, P=0.609 for spike number). The difference between *Pm3b* and Sb lines exceeded the variation among conventional varieties of *T. aestivum* for the traits yield, seed set and vegetative mass (Tables S2 and S3 in Supplementary Information).

The four *Pm3b* lines differed significantly from each other in all the investigated traits (P<0.01 for yield, spike number, seed set, vegetative mass, phenological stage and plant height). Among the four *Pm3b* lines, line *Pm3b*#2 had the lowest yield, seed set and vegetative mass. Seed yield and seed set were reduced by 47% and vegetative mass by 28% compared to the corresponding control line Sb#2. The four control Sb lines slightly differed from each other in vegetative mass (P=0.018) with lower values observed for the lines Sb#2 and Sb#3 than for the other two lines.

Line A9 *Chi*, carrying one copy of the transgene for chitinase production, did not differ from the mother variety Frisal, whereas line A13 *Chi/Glu*, carrying two transgenes for chitinase and glucanase, had significantly lower yield and seed set compared with Frisal and the A9 *Chi* line (P<0.001).

Main effects of environments (soil nutrients and wheat environments)

Nutrient addition enhanced plant growth and development causing an increase in all the investigated traits (P<0.001). Phytometers planted with transgenic *Pm3b* lines as a competitor had higher yield, seed set, spike number and vegetative mass compared with phytometers grown with Sb control lines (P<0.001 for vegetative mass, yield and spike number, P=0.001 for seed set). No differences between *Pm3b* and Sb competitive environments were observed for phenological stage and plant height of phytometers. The variation explained by the contrast between *Pm3b* and Sb control environments exceeded the variation between the three environments of conventional varieties of *T. aestivum* for the phytometer traits yield, spike number, seed set and vegetative mass (Tables S2 and S3). Phytometers which had Frisal as a competitive environment produced less seeds per spike and had delayed phenological development compared with phytometers with Bobwhite as a competitive environment (P=0.039, P=0.004 for seed set and phenological stage, respectively). The phytometers planted in the different Frisal environments (including A9 *Chi* and A13 *Chi/Glu* lines and mother variety) only varied in phenological stage, which was slightly delayed in the transgenic A9 *Chi* and A13 *Chi/Glu* environments (P=0.025).

Overall transgene × environment interactions

Fertilization did not significantly change the range of differences between *Pm3b* and control Sb lines. However, GM and control Frisal phytometers responded differently to nutrient addition (interaction Frisal vs. A9 *Chi* and A13 *Chi/Glu* phytometers × fertilizer: P=0.034 for seed set per spike, P=0.022 for plant height). The difference in seed set between the Frisal mother variety and transgenic lines became smaller but that in height became larger when fertilizer was added to the soil.

Transgenic *Pm3b* lines showed stronger reductions in yield, seed set and vegetative mass compared with corresponding control Sb lines when they were grown in Frisal as opposed to Bobwhite environment (interaction *Pm3b* vs. Sb phytometers × Frisal vs. Bobwhite environment: P=0.001, P=0.006, P=0.034 for yield, seed set and vegetative mass, respectively).

Home vs. away contrast of transgene × environment interactions (relative performance)
In order to assess the performance of transgenic and conventional lines of *T. aestivum* in "away" environments relative to "home" (own) environment, we used the log-ratio away/home as dependent variable. For each phytometer line × fertilizer × environment combination there were four replicate log-ratios according to the four blocks in the field. The overall mean of the log-ratio away/home was positive for the traits yield and vegetative mass ($P<0.001$ for vegetative mass, $P=0.009$ for yield) and negative for height and phenological stage ($P=0.027$ for height, $P=0.001$ for phenological stage; Tables S4 and S5). Bobwhite control lines and conventional Swiss varieties of *T. aestivum* had generally better performance in away than in home environments (Figure 1b). However, Bobwhite *Pm3b* lines had a significantly lower (and even negative) log-ratio compared to the Sb control lines ($P<0.001$ for log-ratios of yield, spike number, seed set and vegetative mass; $P=0.005$, $P=0.012$ for log-ratios of phenological stage and plant height, respectively), indicating a lower performance in away than in home environments. The four *Pm3b* lines differed in their log-ratios of yield, spike number, seed set, plant height and vegetative mass ($P<0.001$): *Pm3b*#2 and *Pm3b*#4 lines suffered most in away as compared with home environments, with *Pm3b*#2 showing reductions of 60% in yield and 50% in seed set and vegetative mass. There were also significant differences between the log-ratios of the four Sb lines ($P<0.001$ for log-ratios of yield, spike number, plant height and phenological stage, $P=0.001$ for log-ratio of seed set, $P=0.026$ for log-ratio of vegetative mass).

Frisal transgenic lines A9 *Chi* and A13 *Chi/Glu* had slightly advanced phenological development and plant height in away compared to home environments, whereas Frisal mother variety was more advanced and taller in home than in away environments (contrast Frisal vs. A9 *Chi* and A13 *Chi/Clu*: $P<0.001$ for log-ratios of plant height and phenological stage). The line A13 *Chi/Glu* had increased vegetative mass in away as compared to home environments, whereas A9 *Chi* line showed no such effect (contrast A9 *Chi* vs. A13 *Chi/Clu*: $P=0.026$ for log-ratio of vegetative mass). Fertilizer addition reduced the log-ratios for yield ($P=0.014$), seed set ($P=0.001$), vegetative mass ($P=0.007$) and phenological stage ($P=0.021$), indicating that line-mixtures may be less beneficial under high than under low soil nutrient conditions.

Discussion

Main effects of transgenes

Our first question was whether introduced transgenes improve plant resistance to mildew and thereby plant performance, which in turn could increase the competitive performance and invasion ability of the GM compared with control plants when the pathogen is present in the environment. As expected, resistance to mildew was increased in the case of the *Pm3b* transgene but not in the case of *Chi* and *Glu* transgenes, presumably because the latter were introduced into the old Swiss wheat variety Frisal which already had a high resistance to mildew in the field. It is conceivable, however, that under higher pathogen pressure the Frisal mother variety might have been more strongly affected by mildew.

In contrast to our expectation, increased mildew resistance did not lead to increased growth and competitive performance of the tested GM lines in the presence of the pathogen. This suggests that the costs were high enough to overcome the benefits of being resistant to powdery mildew, reducing the plants' fitness and their ability to withstand competition from neighbours. Two previous studies have reported similar effects, i.e. increased relative fitness reductions under competition (Heil et al. 2000; van Dam and Baldwin 2001).

These fitness costs also indicate that, at least under the environmental conditions encountered in our field experiment, the mildew-resistant GM lines do not have a higher chance than conventional lines to establish and persist in wheat habitats. On the contrary, the plants with increased mildew resistance due to the presence of a transgene (*Pm3b* lines) had lower yield and reduced seed set than their corresponding control lines and are thus expected to have lower chances to become invasive in the field.

There is a discussion in the literature about what makes a species "weedy" and if already the addition of a single gene can cause a crop to become weedy (Baker 1974; Williamson et al. 1990; Luby and McNicol 1995). Some authors state that weediness arises from many different characters and, therefore, if the species previously had no weedy characteristics, the addition of one or a few genes should not alter its competitiveness (Baker 1974; Luby and McNicol 1995). Our study, however, found that the introduction of a single gene of resistance to a fungal pathogen can have strong effects on the overall plant phenotype (see also Zeller et al. 2010). This supports the point of view that even small genetic changes with the insertion of one gene can cause large ecological alterations affecting plant × environment interactions (Williamson et al.

1990; Williamson 1994; Dale et al. 2002). However, the effects of the transgene in our study were in the direction of decreased rather than increased potential weediness. Interestingly, there were strong differences between the four *Pm3b* lines in their performance and interactions with the biotic and abiotic environments. The line *Pm3b*#2, which showed the highest resistance to powdery mildew, was the weakest competitor, re-enforcing the view that transgene-caused mildew resistance was negatively correlated with plant performance (Figure 2). This line showed a higher level of the transgene expression (see Zeller *et al.* 2010) than the lines *Pm3b*#1, *Pm3b*#3 and *Pm3b*#4, thus indicating that the overexpression of the gene of resistance might be one cause of the changes in the plants' interactions with their environment. Because the corresponding control lines passed through the same transformation procedure as *Pm3b* lines but did not show reduced performance, we assume that the reduced performance in *Pm3b* lines was a consequence of the physiological costs they paid for the increased resistance to the pathogen (Bergelson and Purrington 1996; Brown 2002; Heil and Baldwin 2002).

Line A13 *Chi/Glu*, which possesses both chitinase and glucanase transgenes, had lower yield and seed set than the Frisal variety, without showing an increased resistance (see above). In this case it seems that the additional but unnecessary potential to resist the pathogen, conferred to the GM Frisal plants by the combination of the two transgenes, again was costly for these plants and thus led to decreased growth and performance. Because performance was not reduced in the Frisal line with only one transgene (A9 *Chi*), it again appears that the degree of defence matters for the costs of defence. We conclude that a high "intrinsic" level of mildew resistance has negative effects on the performance of GM plants and thus reduces their potential to persist or even become invasive. It could be suggested that lower levels of intrinsic resistance to pathogens might produce better-performing GM plants. From a risk perspective, however, such plants would have to be evaluated again in a range of biotic and abiotic environments in similar experiments as the one presented here.

Main effects of environments

Our second question was how variation in the abiotic (fertilization) and biotic environment (competition with other wheat lines) may influence resistance to mildew and the performance of phytometer plants. Nutrient addition enhanced powdery mildew incidence in both transgenic and conventional lines of *T. aestivum*. Similar effects were reported in previous studies with non-transgenic plants, where the severity of mildew infection was shown to be related to the nitrogen supply of the host (Last 1953; Bainbridge 1974; Shaner and Finney 1977; Chen et al. 2007). Lines with high mildew incidence proved to be infective biotic environments as shown by the higher mildew incidence of phytometers in these. This is a well-known epidemiological effect (Wolfe 1985) and relevant when considering planting mixed-line crops. However, in the same way as more susceptible neighbours can increase infection in less susceptible target plants so can more resistant neighbours reduce infection in less resistant target plants. In a further field experiment we found that indeed overall mildew incidence in line mixtures was lower than in the average single-line stand (S. Zeller *et al.*, unpublished data), an observation previously made in a genetic diversity experiment with the wild plant species *Solidago canadensis* (Schmid 1994).

Fertilization enhanced plant growth and reproduction in all the investigated lines of *T. aestivum*. In addition, the performance of the phytometer plants depended significantly on the type of competitive environment. Phytometers planted into a competitive environment of transgenic *Pm3b* lines outperformed those planted into a competitive environment of Sb lines. This is in accordance with the results of the analysis of the main effects of transgenes (see previous section). Phytometers which had Frisal variety as a competitive environment generally had weaker performance than those in Bobwhite environments. The congruence between phytometer-line and competitive environment-line effects indicates that phytometers do provide realistic measures of competitive ability.

Overall transgene x environment interactions

The third question asked whether transgenic lines responded to variation in nutrient and competitive environments in the same way as conventional lines of *T. aestivum*, which was the case for mildew infection and for biotic-environment interactions. The difference in mildew incidence between *Pm3b* lines and control increased with the addition of fertilizer. This again has been previously observed with non-transgenic plants, where the increased severity of infection due to fertilization was more pronounced in susceptible than in resistant varieties (Shaner and Finney 1977). In accordance with these observations, the differences in mildew incidence between resistant transgenic lines and control lines also became stronger in mildew-susceptible than in mildew-resistant biotic environments.

Significant transgene × abiotic environment interactions were found only for Frisal lines and only for two traits increasing (for height) or decreasing (for seed set) the differences between the variety Frisal and genetically modified lines A9 *Chi* and A13 *Chi/Glu*. However, significant transgene × biotic environment interactions were observed for the majority of fitness-related traits and reflected more sensitive responses to competition for transgenic Bobwhite than for other lines.

Overall, our findings indicate that a single gene of resistance to a pathogen might strongly affect plant × environment interactions making ecological costs of resistance apparent even in the presence of the pathogen. This, in particular, points to the importance of testing transgenic plants under a set of biotic and abiotic environments in the field.

Relative performance in home vs. away environments

The fourth question asked whether transgenic and non-transgenic lines behave differently if planted into their own (home) rather than into different lines as competitive environments (away). On average, the phytometers benefited if neighbours belonged to a different line (mixture effect). This is consistent with findings in biodiversity experiments (Balvanera et al. 2006).

Whereas the transgenic *Pm3b* lines only performed well in their own competitive environment, all other lines generally showed higher performance in line mixtures. Because the GM line *Pm3b*#2 suffered most in line mixtures, it appears that this line paid a particularly high fitness cost for its elevated mildew resistance level under competition. This supports the recent findings that competition for limited resources might increase the magnitude of fitness costs of resistance (Agrawal 2000; Heil et al. 2000; van Dam and Baldwin 2001). Our results, however, also point to the importance of the type of the competitor and the expression level of the resistance gene. The resistant line with the highest gene expression appeared to be especially sensitive to inter-line competition, whereas the differences between this line and control line became smaller when the competitor was represented by its own genotype.

Our results demonstrate that a transgene increasing plant resistance to a pathogen may reduce rather than increase the plant's competitive ability and thus lower its probability to persist outside its own field. This is in agreement with an early study of Crawley *et al.* (1993) who showed that plants with induced resistance to unfavourable environmental factors were less invasive and persistent than their conventional counterparts.

Apart from these findings, nutrient addition negatively affected the ability of plants to coexist in the mixtures. This supports the theory that fertilization increases competition between genotypes or species for scarce resources and in particular light (Wilson and Tilman 1993; Hautier et al. 2009). It would therefore be even more difficult for competitively weak transgenic plants to invade well fertilized habitats.

The phytometer technique

Here, we have for the first time used an ecological phytometer approach to assess competitive interactions and gene × environment interactions in plants. An advantage of this technique is that many lines can be tested simultaneously on a relatively small area. In our experiment, the performance of 15 different GM and conventional lines of *T. aestivum* was assessed on less than 130 m^2 of field plots. Another advantage of the phytometer technique is the possibility to incorporate many biotic and abiotic factors that might affect the performance and competitive ability of test plants simultaneously into a single and comprehensive experimental design. In our study, for example, we combined 15 biotic environments and two nutrient levels. If transgenic lines would outperform corresponding control lines in one of these realistic field habitats it would be detected by the analysis.

In addition to measuring the competitiveness of the individual phytometer plants, the design of our experiment also allowed assessment of the competitive strength of the environment provided by each line of *T. aestivum*. If the phytometers of different lines on average benefit from being in a certain competitive environment it indicates that the line representing this environment is not a strong competitor.

We believe that the phytometer approach has a high potential in applied ecological and agricultural research. In the future it could be used to facilitate the identification of more or less promising new breeds and increase the flexibility and power of ecological risk assessment.

Conclusions

In conclusion, this study shows that a single gene conferring resistance against a particular fungal pathogen can have large and negative effects on plant performance under realistic field environmental conditions even if these conditions include the presence of the pathogen. We interpret these large costs in resistant plants as a consequence of altered gene regulation, in particular enhanced gene expression level, which was here achieved with a strong promoter introduced with the gene of resistance. This indicates that altered regulation in a single gene may strongly affect plant fitness and the way the plant interacts with the environment, in particular changing a plant's competitive ability.

Acknowledgements

We thank S. Brunner, B. Keller, C. Sautter, J. Fütterer and A. Fammartino for seed material; B. Keller, C. Sautter, W. Gruissem and L. Turnbull for discussions and comments, the national research station Agroscope Reckenholz-Tänikon ART for setting up the field experiment and Y. Kostetskyi and numerous helpers for assistance in the field. This project was supported by the Swiss National Science Foundation and is a part of the wheat-cluster.ch, a sub-unit of the national research programme NRP 59 (SNF 405940-115607).

References

Agrawal, A. A. (2000) Benefits and costs of induced plant defense for *Lepidium virginicum* (*Brassicaceae*). *Ecology,* **81,** 1804–1813.

Allard, R. W. & Adams, J. (1969) Population studies in predominantly self-pollinating species.XIII. Intergenotypic competition and population structure in barley and wheat. *American Naturalist,* **103,** 621–645.

Ammann, K., Jacot, Y. & Al Mazyad, P. R. (2000) Weediness in the light of new transgenic crops and their potential hybrids. *Journal of Plant Diseases and Protection,* **17,** 19–29.

Bainbridge, A. (1974) Effect of nitrogen nutrition of host on barley powdery mildew. *Plant Pathology,* **23,** 160–161.

Baker, H. G. (1974) The evolution of weeds. *Annual Review of Ecology, Evolution, and Systematics,* **5,** 1–24.

Baldwin, I. T. (1988) Damage-induced alkaloids in tobacco – pot-bound plants are not inducible. *Journal of Chemical Ecology,* **14,** 1113–1120.

Balvanera, P., Pfisterer, A. B., Buchmann, N., He, J. S., Nakashizuka, T., Raffaelli, D. & Schmid, B. (2006) Quantifying the evidence for biodiversity effects on ecosystem functioning and services. *Ecology Letters,* **9,** 1146–1156.

Bartsch, D., Schmidt, M., PohlOrf, M., Haag, C. & Schuphan, I. (1996) Competitiveness of transgenic sugar beet resistant to beet necrotic yellow vein virus and potential impact on wild beet populations. *Molecular Ecology,* **5,** 199–205.

Bergelson, J. & Purrington, C. B. (1996) Surveying patterns in the cost of resistance in plants. *American Naturalist,* **148,** 536–558.

Bieri, S., Potrykus, I. & Futterer, J. (2003) Effects of combined expression of antifungal barley seed proteins in transgenic wheat on powdery mildew infection. *Molecular Breeding,* **11,** 37–48.

Brown, J. K. M. (2002) Yield penalties of disease resistance in crops. *Current Opinion in Plant Biology,* **5,** 339–344.

Chen, Y. X., Zhang, F. D., Tang, L., Zheng, Y., Li, Y. J., Christie, P. & Li, L. (2007) Wheat powdery mildew and foliar N concentrations as influenced by N fertilization and belowground interactions with intercropped faba bean. *Plant and Soil,* **291,** 1–13.

Clements, F. E. & Goldsmith, G. W. (1924) *The phytometer method in ecology: the plant and community as instruments.* Carnegie Institution of Washington, Washington.

Crawley, M. J. (1992) The comparative ecology of transgenic and conventional crops. *2nd international symposium on the biosafety results of field tests of genetically modified plants and microorganisms* (eds R. Casper & J. Landsmann), pp. 43–52. Biologische Bundesanstalt fiir Land- und Forstwirtschaft, Berlin, Goslar, Germany.

Crawley, M. J., Hails, R. S., Rees, M., Kohn, D. & Buxton, J. (1993) Ecology of transgenic oilseed rape in natural habitats. *Nature,* **363,** 620–623.

Dale, P. J., Clarke, B. & Fontes, E. M. G. (2002) Potential for the environmental impact of transgenic crops. *Nature Biotechnology,* **20,** 567–575.

Fredshavn, J. R. & Poulsen, G. S. (1996) Growth behavior and competitive ability of transgenic crops. *Field Crops Research,* **45,** 11–18.

Fuchs, M., Chirco, E. M., Mcferson, J. R. & Gonsalves, D. (2004) Comparative fitness of a wild squash species and three generations of hybrids between wild × virus-resistant transgenic squash. *Environmental Biosafety Research,* **3,** 17–28.

Gurr, S. J. & Rushton, P. J. (2005) Engineering plants with increased disease resistance: what are we going to express? *Trends in Biotechnology,* **23,** 275–282.

Hautier, Y., Niklaus, P. A. & Hector, A. (2009) Competition for light causes plant biodiversity loss after eutrophication. *Science,* **324,** 636–638.

Heil, M. (2002) Ecological costs of induced resistance. *Current Opinion in Plant Biology,* **5,** 345–350.

Heil, M. & Baldwin, I. T. (2002) Fitness costs of induced resistance: emerging experimental support for a slippery concept. *Trends in Plant Science,* **7,** 61–67.

Heil, M., Hilpert, A., Kaiser, W. & Linsenmair, K. E. (2000) Reduced growth and seed set following chemical induction of pathogen defence: does systemic acquired resistance (SAR) incur allocation costs? *Journal of Ecology,* **88,** 645–654.

Herms, D. A. & Mattson, W. J. (1992) The dilemma of plants – to grow or defend. *Quarterly Review of Biology,* **67,** 283–335.

Joshi, J., Schmid, B., Caldeira, M. C., Dimitrakopoulos, P. G., Good, J., Harris, R., Hector, A., Huss-Danell, K., Jumpponen, A., Minns, A., Mulder, C. P. H., Pereira, J. S., Prinz, A., Scherer-Lorenzen, M., Siamantziouras, A.-S. D., Terry, A. C., Troumbis, A. Y. & Lawton, J. H. (2001) Local adaptation enhances performance of common plant species. *Ecology Letters,* **4,** 536–544.

Last, F. T. (1953) Some effects of temperature and nitrogen supply on wheat powdery mildew. *Annals of Applied Biology,* **40,** 312–322.

Laughlin, K. D., Power, A. G., Snow, A. A. & Spencer, L. J. (2009) Risk assessment of genetically engineered crops: fitness effects of virus-resistance transgenes in wild *Cucurbita pepo. Ecological Applications,* **19,** 1091–1101.

Leah, R., Tommerup, H., Svendsen, I. & Mundy, J. (1991) Biochemical and molecular characterization of 3 barley seed proteins with antifungal properties. *Journal of Biological Chemistry,* **266,** 1564–1573.

Luby, J. J. & McNicol, R. J. (1995) Gene flow from cultivated to wild raspberries in scotland – developing a basis for risk assessment for testing and deployment of transgenic cultivars. *Theoretical and Applied Genetics,* **90,** 1133–1137.

Mccullagh, P. & Nelder, J. A. (1989) *Generalized linear-models.* Chapman and Hall, London.

McGilchrist, C. A. (1965) Analysis of competition experiments. *Biometrics,* **21,** 975–985.

McGraw, J. B. (1985) Experimental ecology of *Dryas octopetala* ecotypes – relative response to competitors. *New Phytologist,* **100,** 233–241.

Melchers, L. S. & Stuiver, M. H. (2000) Novel genes for disease-resistance breeding. *Current Opinion in Plant Biology,* **3,** 147–152.

Mwangi, P. N., Schmitz, M., Scherber, C., Roscher, C., Schumacher, J., Scherer-Lorenzen, M., Weisser, W. W. & Schmid, B. (2007) Niche pre-emption increases with species richness in experimental plant communities. *Journal of Ecology,* **95,** 65–78.

Rooke, L., Byrne, D. & Salgueiro, S. (2000) Marker gene expression driven by the maize ubiquitin promoter in transgenic wheat. *Annals of Applied Biology,* **136,** 167–172.

Schmid, B. (1994) Effects of genetic diversity in experimental stands of *Solidago altissima* – evidence for the potential role of pathogens as selective agents in plant populations. *Journal of Ecology,* **82,** 165–175.

Shaner, G. & Finney, R. E. (1977) The effect of nitrogen fertilization on the expression of slow-mildewing resistance in Knox wheat. *Purdue University Agricultural Experiment Station Series,* **6408,** 1051–1056.

Strauss, S. Y., Rudgers, J. A., Lau, J. A. & Irwin, R. E. (2002) Direct and ecological costs of resistance to herbivory. *Trends in Ecology & Evolution,* **17,** 278–285.

Tiedje, J. M., Colwell, R. K., Grossman, Y. L., Hodson, R. E., Lenski, R. E., Mack, R. N. & Regal, P. J. (1989) The planned introduction of genetically engineered organisms - ecological considerations and recommendations. *Ecology,* **70,** 298–315.

Tollrian, R. & Harvell, C. D. (1999) The evolution of inducible defenses: current ideas. *The ecology and evolution of inducible defense* (eds R. Tollrian & C. D. Harvell), pp. 306–321. Princeton University Press.

van Dam, N. M. & Baldwin, I. T. (2001) Competition mediates costs of jasmonate-induced defences, nitrogen acquisition and transgenerational plasticity in *Nicotiana attenuata. Functional Ecology,* **15,** 406–415.

van Kleunen, M. & Schmid, B. (2003) No evidence for an evolutionary increased competitive ability in an invasive plant. *Ecology,* **84,** 2816–2823.

Violle, C., Garnier, E., Lecoeur, J., Roumet, C., Podeur, C., Blanchard, A. & Navas, M. L. (2009) Competition, traits and resource depletion in plant communities. *Oecologia,* **160,** 747–755.

Williamson, M. (1994) Community response to transgenic plant release: predictions from British experience of invasive plants and feral crop plants. *Molecular Ecology,* **3,** 75–79.

Williamson, M., Perrins, J. & Fitter, A. (1990) Releasing genetically engineered plants - present proposals and possible hazards. *Trends in Ecology & Evolution,* **5,** 417–419.

Wilson, S. D. & Tilman, D. (1993) Plant competition and resource availability in response to disturbance and fertilization. *Ecology,* **74,** 599–611.

Wolfe, M. S. (1985) The current status and prospects of multiline cultivars and variety mixtures for disease resistance. *Annual Review of Phytopathology,* **23,** 251–273.

Yahiaoui, N., Srichumpa, P., Dudler, R. & Keller, B. (2004) Genome analysis at different ploidy levels allows cloning of the powdery mildew resistance gene *Pm3b* from hexaploid wheat. *Plant Journal,* **37,** 528–538.

Zadoks, J. C., Chang, T. T. & Konzak, C. F. (1974) Decimal code for growth stages of cereals. *Weed Research,* **14,** 415–421.

Zeller, S., Kalinina, O., Brunner, S., Keller, B. & Schmid, B. (2010) Transgene × environment interactions in genetically modified wheat. *PLoS ONE,* **5,** e11405.

Zhu, Q., Maher, E. A., Masoud, S., Dixon, R. A. & Lamb, C. J. (1994) Enhanced protection against fungal attack by constitutive coexpression of chitinase and glucanase genes in transgenic tobacco. *Bio-Technology,* **12,** 807–812.

Figure 5: Photograph taken by Simon Zeller

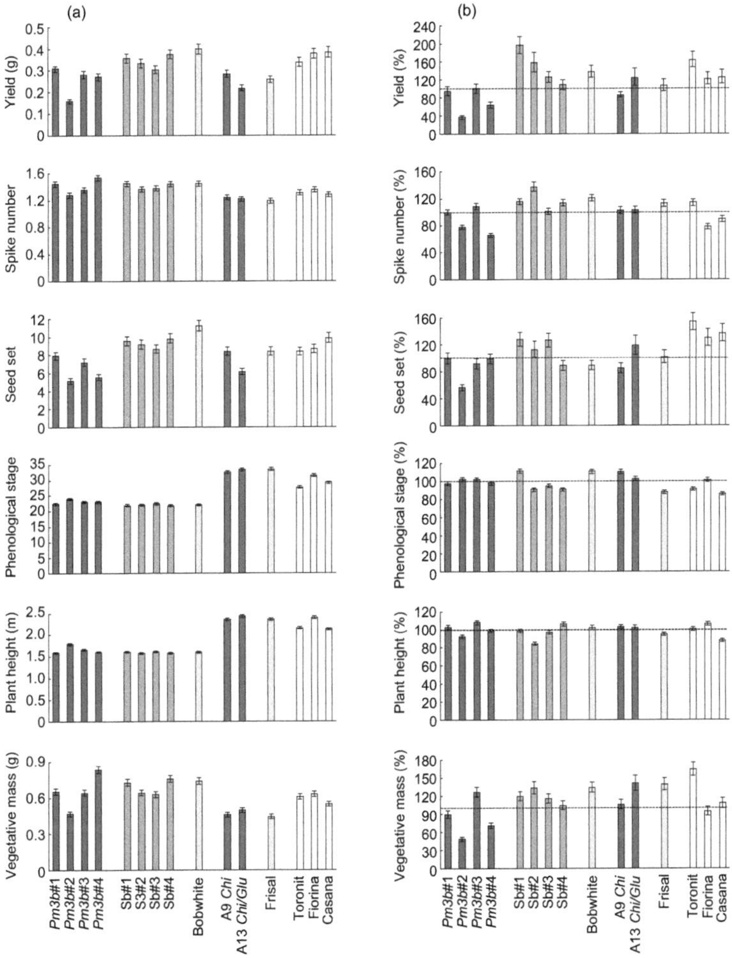

Figure 1. Performance of the 15 transgenic and conventional lines of *T. aestivum* grown with the same lines as competitive environments. Left column (a) gives average line performance across 15 competitive environments. Right column (b) presents relative performance of the investigated wheat lines under competition with other lines and varieties expressed as a percentage of the estimates in their own environment. Dashed line denotes 100% (i.e. log-ratio = 0: same performance in own and foreign competitive environment). Bars show means ± standard errors. Five grades of the grey scale indicate groups of wheat lines; from dark to light: transgenic lines, the genetically closest control (sister lines), wheat varieties used for transgene insertion and modern conventional wheat varieties.

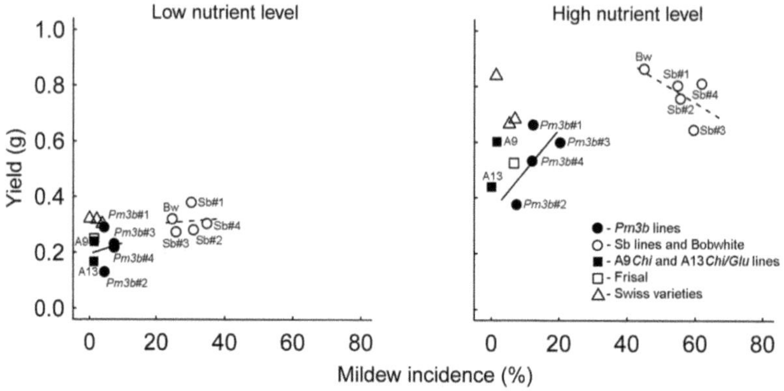

Figure 2. The relationship between *B. graminis* incidence (percentage of plants infected with the pathogen) and yield in 15 lines of *T. aestivum* in two fertilizer treatments. The solid and dotted lines are linear regression lines for the groups of means for transgenic *Pm3b* lines and for control Sb lines and mother variety Bobwhite.

Supplementary Information

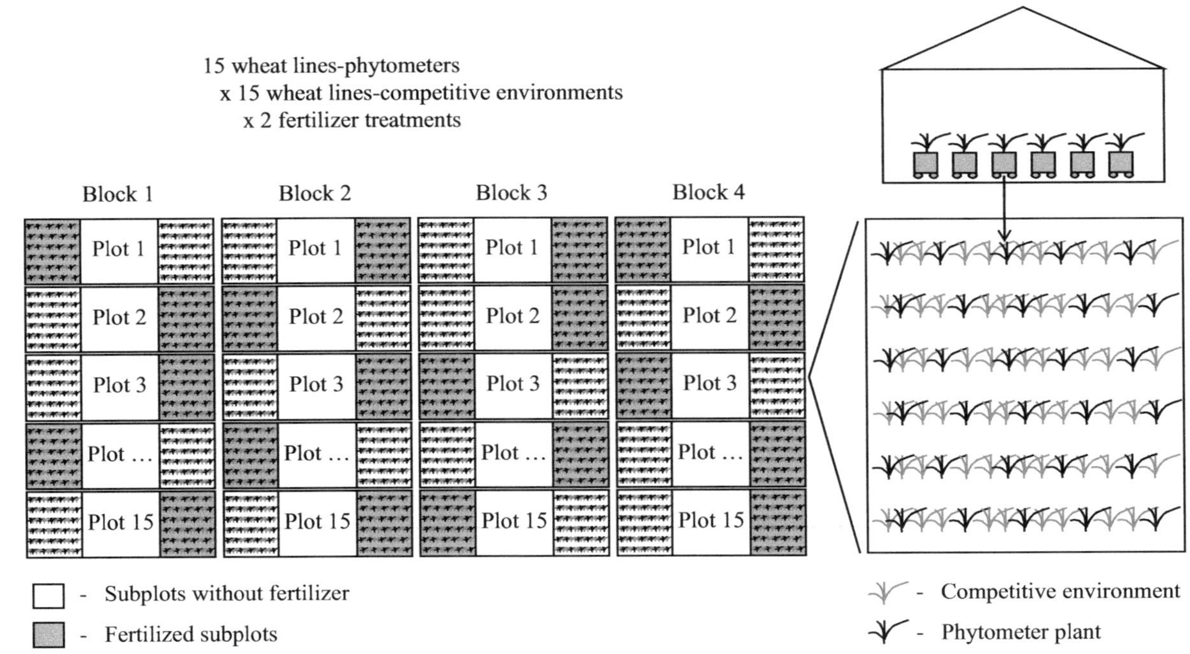

Figure S1: Design of the phytometer experiment.

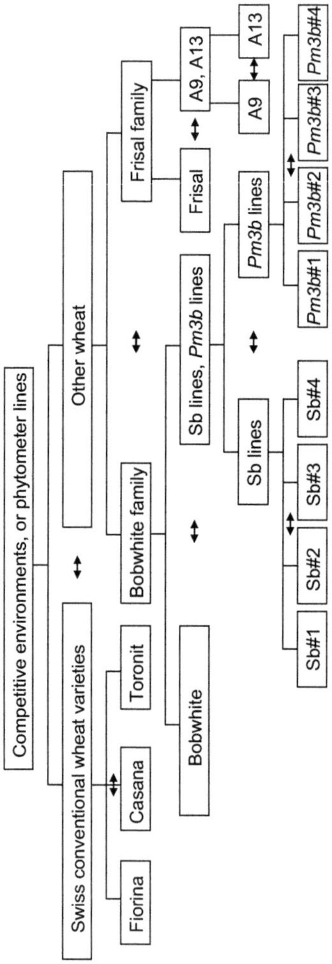

Figure S2: The structure of orthogonal contrasts used in the extended ANOVA models.

Table S1: The analysis of deviance table shows the effects of fertilizer, competitive environment, differences between GM and non-GM lines and their interactions on mildew incidence.

Simple model

Source of variation	df	%SS	F pr.
Block	3	1.1	0.018
Competitive environment (Comp.env.)	14	5.2	0.001
Plot	42	4.3	0.040
Fertilizer	1	3.1	<.001
Comp.env.×Fertilizer	14	1.2	0.202
Subplot	45	2.7	0.949
Phytometer lines	14	26.8	<.001
Comp.env.×Phytometer lines	196	7.1	0.067
Plot×Phytometer lines	593	18.2	<.001
Phytometer lines×Fertilizer	14	1.2	<.001
Residual	1513	29.0	
Total	2449	100	

Extended model

Source of variation	df	%SS	F pr.
Block	3	1.1	0.018
Competitive environment (Comp.env.)	14	5.2	0.001
Plot	42	4.3	0.040
Fertilizer	1	3.1	<.001
Comp.env.×Fertilizer	14	1.2	0.202
Subplot	45	2.7	<.001
Phytometer contrasts (Phytometer lines effect):			
Swiss vs. other wheat	1	5.7	<.001
3 conventional Swiss varieties	2	0.3	<.001
Bobwhite vs. Frisal	1	7.6	<.001
Bobwhite vs. Sb lines	1	0.3	<.001
Sb lines vs. *Pm3b* lines	1	12.4	<.001
4 Sb lines	3	0.1	0.059
4 *Pm3b* lines	3	0.3	0.001
Frisal vs. A9 and A13	1	0.1	0.008
A9 vs. A13	1	0.0	0.331
Comp.env.×Swiss vs. other wheat	14	0.4	0.069
Comp.env.×3 conventional Swiss varieties	28	0.7	0.671
Comp.env.×Bobwhite vs. Frisal	14	0.7	0.088
Comp.env.×Bobwhite vs. Sb lines	14	0.3	0.728
Comp.env.×Sb lines vs. *Pm3b* lines	14	0.8	0.024
Comp.env.×4 Sb lines	42	2.0	0.014
Comp.env.×4 *Pm3b* lines	42	1.6	0.138
Comp.env.×Frisal vs. A9 and A13	14	0.3	0.756
Comp.env.×A9 vs. A13	14	0.2	0.947
Plot×Phytometer lines	593	18.2	<.001
Fertilizer×Swiss vs. other wheat	1	0.1	0.119
Fertilizer×3 conventional Swiss varieties	2	0.0	0.487
Fertilizer×Bobwhite vs. Frisal	1	0.0	0.758
Fertilizer×Bobwhite vs. Sb lines	1	0.0	0.843
Fertilizer×Sb lines vs. *Pm3b* lines	1	0.2	<.001
Fertilizer×4 Sb lines	3	0.3	0.002
Fertilizer×4 *Pm3b* lines	3	0.1	0.225
Fertilizer×Frisal vs. A9 and A13	1	0.2	0.002
Fertilizer×A9 vs. A13	1	0.4	<.001
Residual	1513	29.0	
Total	2449	100	

Table S2: The ANOVA table shows the effects of fertilizer, competitive environment, differences between GM and non-GM lines and their interactions on three yield characteristics.

Simple model	Yield (log)			Spike number (log)			Seed set (log)		
Source of variation	df	%SS	F pr.	df	%SS	F pr.	df	%SS	F pr.
Block	3	4.2	<.001	3	2.9	<.001	3	1.8	0.003
Competitive environment	14	8.3	<.001	14	5.9	<.001	14	5.2	0.001
Plot	42	6.4	<.001	42	3.4	0.376	42	4.6	0.001
Fertilizer	1	17.2	<.001	1	9.7	<.001	1	9.3	<.001
Comp.env.×Fertilizer	14	0.5	0.596	14	0.8	0.709	14	0.8	0.224
Subplot	45	1.9	0.007	45	3.3	<.001	45	1.9	0.068
Phytometer lines	14	5.9	<.001	14	2.0	<.001	14	7.8	<.001
Comp.env.×Phytometer lines	196	5.4	0.071	196	6.1	0.583	196	6.7	0.075
Plot×Phytometer lines	593	13.9	0.888	598	19.2	0.171	593	17.2	0.924
Phytometer lines×Fertilizer	14	0.2	0.896	14	0.7	0.085	14	0.4	0.518
Residual	1406	36.0		1522	45.9		1387	44.4	
Total	2342	100		2463	100		2323	100	

Extended model									
Source of variation	df	%SS	F pr.	df	%SS	F pr.	df	%SS	F pr.
Block	3	4.2	<.001	3	2.9	<.001	3	1.8	0.003
Competitive environment contrasts:									
Swiss vs. other wheat	1	1.7	0.002	1	0.7	0.006	1	1.2	0.002
3 conventional Swiss varieties	2	0.5	0.183	2	0.9	0.009	2	0.2	0.419
Bobwhite vs. Frisal	1	0.4	0.137	1	0.0	0.679	1	0.5	0.039
Bobwhite vs. Sb lines	1	0.0	0.959	1	0.1	0.182	1	0.0	0.719
Sb lines vs. *Pm3b* lines	1	2.7	<.001	1	2.4	<.001	1	1.4	0.001
4 Sb lines	3	1.3	0.044	3	0.1	0.724	3	1.4	0.011
4 *Pm3b* lines	3	1.3	0.047	3	1.7	0.001	3	0.3	0.421
Frisal vs. A9 and A13	1	0.0	0.814	1	0.0	0.909	1	0.0	0.957
A9 vs. A13	1	0.3	0.149	1	0.0	0.841	1	0.2	0.166
Plot	42	6.4	<.001	42	3.4	0.376	42	4.6	0.001
Fertilizer	1	17.2	<.001	1	9.7	<.001	1	9.3	<.001
Comp.env.×Fertilizer	14	0.5	0.596	14	0.8	0.709	14	0.8	0.224
Subplot	45	1.8	0.007	45	3.3	<.001	45	1.9	0.067
Phytometer contrasts:									
Swiss vs. other wheat	1	0.9	<.001	1	0.1	0.073	1	0.4	<.001
3 conventional Swiss varieties	2	0.1	0.204	2	0.0	0.658	2	0.2	0.061
Bobwhite vs. Frisal	1	0.3	<.001	1	1.1	<.001	1	0.1	0.175
Bobwhite vs. Sb lines	1	0.6	<.001	1	0.0	0.459	1	1.4	<.001
Sb lines vs. *Pm3b* lines	1	1.8	<.001	1	0.0	0.609	1	3.7	<.001
4 Sb lines	3	0.1	0.129	3	0.1	0.372	3	0.1	0.515
4 *Pm3b* lines	3	1.8	<.001	3	0.7	<.001	3	1.3	<.001
Frisal vs. A9 and A13	1	0.0	0.441	1	0.0	0.647	1	0.2	0.026
A9 vs. A13	1	0.3	<.001	1	0.0	0.759	1	0.5	<.001
Comp.env.×Phytometer lines	196	5.4	0.071	196	6.1	0.583	196	6.7	0.075
Plot×Phytometer lines	593	13.9	0.888	598	19.2	0.171	593	17.2	0.924
Fertilizer×Swiss vs. other wheat	1	0.0	0.946	1	0.0	0.287	1	0.0	0.356
Fertilizer×3 conventional Swiss var.	2	0.0	0.964	2	0.1	0.438	2	0.0	0.83
Fertilizer×Bobwhite vs. Frisal	1	0.1	0.107	1	0.1	0.062	1	0.0	0.507
Fertilizer×Bobwhite vs. Sb lines	1	0.0	0.781	1	0.0	0.236	1	0.0	0.513
Fertilizer×Sb lines vs. *Pm3b* lines	1	0.0	0.415	1	0.0	0.503	1	0.0	0.371
Fertilizer×4 Sb lines	3	0.0	0.923	3	0.2	0.096	3	0.0	0.913
Fertilizer×4 *Pm3b* lines	3	0.1	0.486	3	0.1	0.613	3	0.2	0.159
Fertilizer×Frisal vs. A9 and A13	1	0.0	0.253	1	0.1	0.128	1	0.1	0.034
Fertilizer×A9 vs. A13	1	0.0	0.642	1	0.1	0.078	1	0.0	0.931
Residual	1406	36.0		1522	45.9		1387	44.4	
Total	2342	100		2463	100		2323	100	

Table S3: The ANOVA table shows the effects of fertilizer, competitive environment, differences between GM and non-GM lines and their interactions on the phenological stage, plant height and vegetative mass.

Simple model	Vegetative mass (log)			Plant height (log)			Phenological stage (log)		
Source of variation	df	%SS	F pr.	df	%SS	F pr.	df	%SS	F pr.
Block	3	2.2	0.005	3	31.8	<.001	3	28.1	<.001
Competitive environment	14	10.1	<.001	14	0.6	0.537	14	1.0	0.136
Plot	42	6.5	<.001	42	2.1	0.030	42	2.0	0.203
Fertilizer	1	27.7	<.001	1	7.0	<.001	1	5.1	<.001
Comp.env.×Fertilizer	14	0.4	0.904	14	0.9	0.017	14	0.4	0.674
Subplot	45	2.2	<.001	45	1.3	<.001	45	1.6	<.001
Phytometer lines	14	5.8	<.001	14	24.2	<.001	14	20.4	<.001
Comp.env.×Phytometer lines	196	3.6	0.896	196	2.0	0.846	196	2.4	0.999
Plot×Phytometer lines	597	12.7	0.082	630	7.4	<.001	630	11.8	<.001
Phytometer lines×Fertilizer	14	0.4	0.160	14	0.6	<.001	14	0.4	0.001
Residual	1473	28.5		2625	22.2		2594	26.9	
Total	2413	100		3598	100		3567	100	

Extended model									
Source of variation	df	%SS	F pr.	df	%SS	F pr.	df	%SS	F pr.
Block	3	2.2	0.005	3	31.8	<.001	3	28.1	<.001
Competitive environment contrasts:									
Swiss vs. other wheat	1	2.4	<.001	1	0.1	0.210	1	0.0	0.691
3 conventional Swiss varieties	2	1	0.050	2	0.1	0.489	2	0.2	0.089
Bobwhite vs. Frisal	1	0.3	0.175	1	0.1	0.200	1	0.4	0.004
Bobwhite vs. Sb lines	1	0.1	0.374	1	0.0	0.604	1	0.0	0.35
Sb lines vs. *Pm3b* lines	1	2.6	<.001	1	0.0	0.969	1	0.0	0.692
4 Sb lines	3	1.7	0.018	3	0.1	0.652	3	0.0	0.998
4 *Pm3b* lines	3	1.7	0.017	3	0.1	0.402	3	0.0	0.878
Frisal vs. A9 and A13	1	0.0	0.703	1	0.1	0.115	1	0.3	0.025
A9 vs. A13	1	0.2	0.297	1	0.0	0.399	1	0.0	0.921
Plot	42	6.4	<.001	42	2.1	0.030	42	2.0	0.203
Fertilizer	1	27.7	<.001	1	7.0	0.000	1	5.1	<.001
Comp.env.×Fertilizer	14	0.4	0.904	14	0.9	0.017	14	0.4	0.674
Subplot	45	2.2	<.001	45	1.3	<.001	45	1.6	<.001
Phytometer contrasts:									
Swiss vs. other wheat	1	0.0	0.395	1	6.0	<.001	1	3.3	<.001
3 conventional Swiss varieties	2	0.1	0.169	2	0.4	<.001	2	0.4	<.001
Bobwhite vs. Frisal	1	2.8	<.001	1	17.1	<.001	1	16.2	<.001
Bobwhite vs. Sb lines	1	0.1	0.013	1	0.0	0.224	1	0.0	0.111
Sb lines vs. *Pm3b* lines	1	0.2	<.001	1	0.2	<.001	1	0.2	<.001
4 Sb lines	3	0.3	<.001	3	0.0	0.515	3	0.0	0.591
4 *Pm3b* lines	3	2.2	<.001	3	0.4	<.001	3	0.1	0.023
Frisal vs. A9 and A13	1	0.0	0.381	1	0.0	0.241	1	0.0	0.508
A9 vs. A13	1	0.0	0.245	1	0.0	0.088	1	0.0	0.195
Comp.env.×Phytometer lines	196	3.6	0.896	196	2.0	0.846	196	2.4	0.999
Plot×Phytometer lines	597	12.7	0.082	630	7.4	<.001	630	11.8	<.001
Fertilizer×Swiss vs. other wheat	1	0.1	0.114	1	0.1	<.001	1	0.0	0.892
Fertilizer×3 conventional Swiss var.	2	0.2	0.019	2	0.0	0.603	2	0.1	0.009
Fertilizer×Bobwhite vs. Frisal	1	0.1	0.095	1	0.3	<.001	1	0.1	0.012
Fertilizer×Bobwhite vs. Sb lines	1	0.0	0.694	1	0.0	0.842	1	0.0	0.196
Fertilizer×Sb lines vs. *Pm3b* lines	1	0.0	0.834	1	0.0	0.830	1	0.0	0.468
Fertilizer×4 Sb lines	3	0.0	0.820	3	0.1	0.024	3	0.0	0.443
Fertilizer×4 *Pm3b* lines	3	0.1	0.359	3	0.1	0.061	3	0.1	0.132
Fertilizer×Frisal vs. A9 and A13	1	0.0	0.426	1	0.0	0.022	1	0.0	0.214
Fertilizer×A9 vs. A13	1	0.0	0.319	1	0.0	0.503	1	0.1	0.004
Residual	1473	28.5		2625	22.2		2594	26.9	
Total	2413	100		3598	100		3567	100	

Table S4: The ANOVA table shows the effects of fertilizer, competitive environment, differences between GM and non-GM lines and their interactions on three relative yield characteristics.

Simple model	Yield (log)			Spike number (log)			Seed set (log)		
Source of variation	df	%SS	F pr.	df	%SS	F pr.	df	%SS	F pr.
Overall mean	1	0.4	0.009	1	0.0	0.503	1	0.1	0.149
Block	3	1.4	0.021	3	1.9	0.001	3	2.0	0.002
Competitive environment	14	7.5	<.001	14	5.9	<.001	14	5.2	0.001
Plot	47	5.9	0.075	47	4.3	0.287	47	5.3	0.121
Fertilizer	1	0.5	0.014	1	0.1	0.290	1	1.1	0.001
Comp.env.×Fertilizer	14	0.6	0.917	14	0.3	0.991	14	0.8	0.732
Subplot	43	3.5	0.059	44	3.4	0.007	43	3.4	0.118
Phytometer lines	14	12.7	<.001	14	13.5	<.001	14	7.8	<.001
Comp.env.×Phytometer lines	180	6.7	0.999	180	6.8	0.999	180	8.9	0.999
Plot×Phytometer lines	502	35.6	0.029	520	40.6	<.001	495	42.0	0.001
Phytometer lines×Fertilizer	13	3.9	<.001	14	3.4	<.001	13	2.3	0.001
Residual	363	21.3		421	19.8		347	21.2	
Total	1195	100		1273	100		1172	100	

Extended model									
Source of variation	df	%SS	F pr.	df	%SS	F pr.	df	%SS	F pr.
Overall mean	1	0.4	0.009	1	0.0	0.503	1	0.1	0.149
Block	3	1.4	0.021	3	1.9	0.001	3	2.0	0.002
Competitive environment	14	7.5	<.001	14	5.9	<.001	14	5.2	0.001
Plot	47	5.9	0.075	47	4.3	0.287	47	5.3	0.121
Fertilizer	1	0.5	0.014	1	0.1	0.290	1	1.1	0.001
Comp.env.×Fertilizer	14	0.6	0.917	14	0.3	0.991	14	0.8	0.732
Subplot	43	3.5	0.059	44	3.4	0.007	43	3.4	0.118
Phytometer contrasts:									
Swiss vs. other wheat	1	1.1	<.001	1	0.2	0.022	1	2.7	<.001
3 conventional Swiss varieties	2	0.2	0.219	2	1.3	<.001	2	0.1	0.346
Bobwhite vs. Frisal	1	0.0	0.447	1	0.1	0.084	1	0.0	0.556
Bobwhite vs. Sb lines	1	0.8	<.001	1	1.2	<.001	1	0.0	0.476
Sb lines vs. *Pm3b* lines	1	5.7	<.001	1	4.4	<.001	1	1.4	<.001
4 Sb lines	3	1.3	<.001	3	1.5	<.001	3	1.1	0.001
4 *Pm3b* lines	3	3.5	<.001	3	4.7	<.001	3	2.3	<.001
Frisal vs. A9 and A13	1	0.0	0.543	1	0.0	0.396	1	0.0	0.435
A9 vs. A13	1	0.1	0.240	1	0.0	0.533	1	0.1	0.326
Comp.env.×Phytometer lines	180	6.7	0.999	180	6.8	0.999	180	8.9	0.999
Plot×Phytometer lines	502	35.6	0.029	520	40.6	<.001	495	42.0	0.001
Fertilizer×Swiss vs. other wheat	1	0.1	0.188	1	0.1	0.100	1	0.2	0.115
Fertilizer×3 conventional Swiss var.	2	0.6	0.009	2	0.2	0.091	2	0.4	0.051
Fertilizer×Bobwhite vs. Frisal	1	0.1	0.257	1	0.0	0.340	1	0.0	0.428
Fertilizer×Bobwhite vs. Sb lines	1	0.6	0.001	1	0.0	0.647	1	0.6	0.002
Fertilizer×Sb lines vs. *Pm3b* lines	1	1.3	<.001	1	1.0	<.001	1	0.0	0.914
Fertilizer×4 Sb lines	2	0.6	0.005	3	0.7	0.002	2	0.2	0.281
Fertilizer×4 *Pm3b* lines	3	0.5	0.033	3	0.5	0.010	3	0.7	0.010
Fertilizer×Frisal vs. A9 and A13	1	0.0	0.731	1	0.0	0.828	1	0.1	0.247
Fertilizer×A9 vs. A13	1	0.0	0.803	1	0.7	<.001	1	0.2	0.091
Residual	363	21.3		421	19.8		347	21.2	
Total	1195	100		1273	100		1172	100	

Table S5: The ANOVA table shows the effects of fertilizer, competitive environment, differences between GM and non-GM lines and their interactions on the relative phenological stage, plant height and vegetative mass.

Simple model	Vegetative mass (log)			Plant height (log)			Phenological stage (log)		
Source of variation	df	%SS	F pr.	df	%SS	F pr.	df	%SS	F pr.
Overall mean	1	0.6	<.001	1	0.2	0.027	1	0.5	0.001
Block	3	0.4	0.439	3	1.0	0.045	3	0.2	0.619
Competitive environment	14	9.5	<.001	14	2.1	0.217	14	2.2	0.259
Plot	47	7.2	0.013	47	5.2	0.723	47	5.9	0.005
Fertilizer	1	0.6	0.007	1	0.0	0.730	1	0.3	0.021
Comp.env.×Fertilizer	14	0.3	0.994	14	1.2	0.823	14	1.4	0.085
Subplot	44	3.4	0.015	44	5.8	<.001	44	2.5	0.144
Phytometer lines	14	13.5	<.001	14	7.7	<.001	14	10.2	<.001
Comp.env.×Phytometer lines	180	5.2	0.999	180	5.5	0.999	180	4.9	0.999
Plot×Phytometer lines	515	35.9	<.001	610	33.3	0.025	610	32.4	0.040
Phytometer lines×Fertilizer	14	3.3	<.001	14	3.5	<.001	14	5.6	<.001
Residual	403	20.1		735	34.5		731	33.9	
Total	1250	100		1677	100		1673	100	

Extended model									
Source of variation	df	%SS	F pr.	df	%SS	F pr.	df	%SS	F pr.
Overall mean	1	0.6	<.001	1	0.2	0.027	1	0.5	0.001
Block	3	0.4	0.439	3	1.0	0.045	3	0.2	0.619
Competitive environment	14	9.5	<.001	14	2.1	0.217	14	2.2	0.259
Plot	47	7.2	0.013	47	5.2	0.723	47	5.9	0.005
Fertilizer	1	0.6	0.007	1	0.0	0.730	1	0.3	0.021
Comp.env.×Fertilizer	14	0.3	0.994	14	1.2	0.823	14	1.4	0.085
Subplot	44	3.4	0.015	44	5.8	<.001	44	2.5	0.144
Phytometer contrasts:									
Swiss vs. other wheat	1	0.7	<.001	1	0.0	0.336	1	1.6	<.001
3 conventional Swiss varieties	2	1.4	<.001	2	2.1	<.001	2	1.5	<.001
Bobwhite vs. Frisal	1	0.7	<.001	1	0.0	0.617	1	0.0	0.685
Bobwhite vs. Sb lines	1	1.2	<.001	1	0.2	0.066	1	1.3	<.001
Sb lines vs. *Pm3b* lines	1	3.3	<.001	1	0.4	0.005	1	0.3	0.012
4 Sb lines	3	0.5	0.026	3	2.9	<.001	3	3.0	<.001
4 *Pm3b* lines	3	5.4	<.001	3	1.4	<.001	3	0.2	0.295
Frisal vs. A9 and A13	1	0.0	0.325	1	0.6	<.001	1	2.0	<.001
A9 vs. A13	1	0.2	0.026	1	0.0	0.832	1	0.3	0.013
Comp.env.×Phytometer lines	180	5.2	0.999	180	5.5	0.999	180	4.9	0.999
Plot×Phytometer lines	515	35.9	<.001	610	33.3	0.025	610	32.4	0.040
Fertilizer×Swiss vs. other wheat	1	0.1	0.226	1	0.0	0.781	1	0.2	0.030
Fertilizer×3 conventional Swiss var.	2	0.0	0.907	2	0.6	0.001	2	0.1	0.373
Fertilizer×Bobwhite vs. Frisal	1	0.0	0.716	1	0.3	0.013	1	0.1	0.235
Fertilizer×Bobwhite vs. Sb lines	1	0.8	<.001	1	0.0	0.544	1	0.6	<.001
Fertilizer×Sb lines vs. *Pm3b* lines	1	1.2	<.001	1	0.6	<.001	1	1.0	<.001
Fertilizer×4 Sb lines	3	1.0	<.001	3	0.1	0.631	3	0.4	0.025
Fertilizer×4 *Pm3b* lines	3	0.1	0.611	3	0.8	0.001	3	0.7	0.002
Fertilizer×Frisal vs. A9 and A13	1	0.1	0.199	1	0.0	0.468	1	0.0	0.320
Fertilizer×A9 vs. A13	1	0.0	0.324	1	1.1	<.001	1	2.4	<.001
Residual	403	20.1		735	34.5		731	33.9	
Total	1250	100		1677	100		1673	100	

CHAPTER 5

Gene flow in genetically modified wheat

S. Rieben, S.L. Zeller, O. Kalinina, B. Schmid, *Manuscript*

Fig. 6: Spikelet of a GM-Bobwhite line (*Pm3b*#2) with non-fertilized flowers

Abstract

Understanding gene flow in genetically modified (GM) crops is critical to answer questions of coexistence of GM and non-GM crops. We tested in two field experiments whether rates of cross-pollination differed between GM and non-GM lines of bread wheat *Triticum aestivum*. In the first experiment, cross-pollination was studied within the field by planting "phytometers" of one line (pollen recipient) into stands of another line (pollen donor). In the second experiment, cross-pollination was studied over distances of 0.5–2.5 m from a central patch of pollen donors to adjacent patches of pollen recipients on opposite sides. Cross-pollination was detected when offspring of a pollen recipient without a particular transgene contained this transgene in heterozygous condition (tested using phenotypic characteristics and a novel population-level molecular method). The GM lines had been produced from the varieties Bobwhite or Frisal and contained *Pm3b* or *chi/glu* transgenes, respectively, in homozygous condition. These transgenes increase plant resistance against pathogenic fungi. Although the overall rate of cross-pollination in the first experiment was only 3.4%, GM lines containing the *Pm3b* transgene were six times more likely to receive foreign pollen than non-GM control lines. Furthermore, the GM line containing a *chi* transgene was a better pollen donor than non-GM lines. One GM line (*Pm3b*#2) showed reduced fertility and higher offspring mortality under herbivore attack. In the second experiment, cross-pollination declined from 0.7–0.03% over the test distances of 0.5–2.5 m and this was independent of prevailing wind direction. Our results suggest that pollen-mediated gene flow between GM and non-GM wheat might only be a concern if it occurs within fields, e.g. due to seed contamination.

Introduction

The frequent use of genetically modified (GM) plants in agriculture demands in-depth ecological risk assessment (Wolfenbarger and Phifer 2000; Cellini et al. 2004; Conner et al. 2003; Snow et al. 2005; Andow and Zwahlen 2006). A possible consequence of the release of GM crops can be unintended gene flow into traditional varieties or wild relatives (Jørgensen and Andersen 1994; Daniell 2002; Rieger et al. 2002; Mercer and Wainwright 2008; Piñeyro-Nelson et al. 2009). Gene flow can increase the ability of a population to respond to a changing environment due to an increased genetic diversity (Gustafson et al. 2005). In plants, gene flow does not only occur by migrating individuals (seed dispersal) but also by migrating gametes, i.e. pollen dispersal. Gene flow via pollen dispersal can occur within and between populations and occasionally even between species (Levin and Kerster 1974; Hedrick 2004). Understanding this process is critical to ensuring the coexistence of GM and non-GM crops (Weber et al. 2007; Pla et al. 2006) and data about pollen-mediated gene flow are essential to establish appropriate isolation distances between traditional varieties and GM plants (Waines and Hegde 2003). These isolation distances should help to achieve the European Union (EU) 0.9% GM-adventitious-presence-labelling threshold for food and feed (Beckie and Hall 2008). Furthermore, gene flow from such adventitious GM plants to non-GM neighbours within a field should be known to predict the maximum level of contamination expected in the offspring seed population given a particular contamination level at sowing.

Previous studies about gene flow in conventional wheat, a strongly selfing species (De Vries 1971), found cross-pollination rates of 1–2% for plants in close proximity (De Vries 1974; Griffin 1987; Martin 1990), which rapidly decreased with greater distance between pollen donor and pollen recipient (De Vries 1971; Gatford et al. 2006). There are several reasons why wheat has such a low cross-pollination rate compared to other grain species. First, fertilization usually occurs before the florets open, which makes pollination with foreign pollen unlikely. Second, although wheat is a wind-pollinated species (Eastham and Sweet 2002), its pollen is relatively heavy and settles quickly compared to other grass species (De Vries 1971). Despite the low rates of gene flow, a maximum cross-pollination distance of 2.75 km has been reported in the literature (Matus-Cádiz et al. 2007). While there are numerous studies about gene flow over certain distances, gene flow within stands of crop plants, including wheat, has rarely been analysed. However, such studies are necessary to assess the potential dispersal of GM traits if GM plants occur as contamination within fields planted with non-GM crops, due to contaminated seed material or volunteer seedlings (Graziano et al. 2007). It is usually assumed that GM-wheat would behave similar to conventional varieties but there is only little evidence corroborating this standpoint (Gatford et al. 2006).

In the present study we used GM spring wheat (*Triticum aestivum* L.) with transgenes conferring resistance against fungal pathogens as a model system to assess gene flow by cross-pollination within stands and over short distances in two field experiments. Different GM lines had been produced from the varieties Bobwhite or Frisal and contained *Pm3b* or *chi/glu* transgenes, respectively, in homozygous condition. *Pm3b* induces constitutive resistance against a specific strain of powdery mildew *Blumeria graminis* f.sp. *tritici* (Yahiaoui et al. 2004) and *chi/glu* had been shown to have quantitative antifungal properties (Bieri et al. 2003). To assess gene flow within the field, we planted seedlings of four independently transformed *Pm3b* and corresponding non-GM control lines as "phytometers" (Clements and Goldsmith 1924; Zeller et al. 2010, Kalinina et al.2011) into plots with four different wheat varieties. These so-called backgrounds consisted of two GM lines (A9 *chi* and A13 *chi/glu*) and two non-GM control lines (Frisal, Casana). Seeds of phytometers that flowered simultaneously with their background were collected and planted in a field trial. Hybrids were then identified by phenotypic traits specific to outcrosses between plants of the variety Bobwhite with Frisal or Casana (hybrid necrosis). To assess short distance gene flow, we grew 2.5 m strips of non-GM control lines east and west of 1 x 1 m GM wheat plots. We determined the cross-pollination rate by pooling offspring seeds from the control lines and testing them for the presence or absence of resistance genes using population-level molecular analyses. The aims of the study were to (i) measure gene flow within the field from two GM and two non-GM lines planted as pollen-donor backgrounds to four different pairs of GM/non-GM sister lines planted as pollen-recipient phytometers, (ii) to measure the influence of distance between GM pollen donor and non-GM pollen recipient on the cross-pollination rates of three pairs of GM/non-GM sister lines and (iii) to analyse line specific differences in rates of cross-pollination.

Material and Methods

Genetically modified wheat

We used six GM lines of spring wheat either derived from the Mexican variety Bobwhite SH 98 26 or the Swiss variety Frisal. Four GM lines from the variety Bobwhite SH 98 26 were produced by biolistic transformation in different transformation events and each line carried a single copy of the transgene *Pm3b* (Zeller et al. 2010). *Pm3b* confers race-specific resistance to powdery mildew and was cloned from hexaploid wheat (Yahiaoui et al. 2004). The transgene was cloned under the control of the maize *Zea mays* L. ubiquitin promoter (Christensen and Quail 1996) and transformants were selected on mannose-containing media using the phosphomannose isomerase (PMI) coding gene as a selectable marker (Reed et al. 2001). After regeneration of T0 transformants, four independent T1 families were selected. From each T1 family, an offspring pair was further propagated consisting of a homozygous GM plant (GM lines *Pm3b*#1–4) and a null-segregant, i.e. a plant that did neither inherit the *Pm3b* transgene nor the selectable marker (control lines S3b#1–4; Zeller et al. 2010).

Two GM lines of the variety Frisal were produced by biolistic transformation using the plasmid MAGUCUM, containing (1) an actin-1 promoter, barley-seed ß-1,3-glucanase (glu) and CaMV terminator, (2) an ubiquitin-1 promoter, barley-seed chitinase (*chi*), CaMV terminator and (3) the bar gene for selection (Bliffeld et al. 1999). The GM line A9 *chi* was positively selected for chitinase expression and the line A13 *chi/glu* for chitinase and glucanase expression (Bieri et al. 2003). The pathogenesis-related proteins chitinase and glucanase are known for their broad antifungal effect and its expression should lead to an increased resistance to powdery mildew (Leah et al. 1991; Zhu et al. 1994).

For the field experiments we used seeds obtained from the fifth generation of the GM lines *Pm3b*#1–4 and their respective non-GM sister lines Sb#1–4 as controls, and seed obtained from the sixth generation of the GM lines A9 *chi* and A13 *chi/glu* and its cultivar Frisal as a control. In addition we used the conventional wheat variety Casana as a further non-GM control line.

Experiment 1: gene flow within plots

The first part of experiment 1 was running from March 2008 until August 2008 (Zeller et al. 2010). It was carried out as part of a lager field trial at an agricultural research station in Zurich-Reckenholz, Switzerland. Seeds of the variety Frisal, its GM lines A9 *chi* and A13 *chi/glu*, and the variety Casana, were sown into eight 1 x 1.08 m plots per line. These stands acted as pollen-donating wheat backgrounds. In each plot, 400 seeds were sown in six rows with a distance of 18 cm between rows using an Oyjord plot drill system (Wintersteiger AG, Ried, Austria). At the same time, seedlings of the four *Pm3b* lines and the four corresponding controls (all variety Bobwhite) were raised individually in the glasshouse and transplanted as "phytometers" into each of the 32 field plots (2 phytometers per line per plot) once they reached the phenological stages 12–13 (Zadoks et al. 1974). These Bobwhite phytometers were therefore surrounded by Frisal or Casana background plants. With this planting procedure we hoped to maximize chances for pollen transfer from background to phytometer plants. Furthermore, it allowed us to detect outcrossed offspring later on because hybrids between Frisal or Casana and Bobwhite differ morphologically from the parental varieties. The flowering period of background and phytometer plants was continuously recorded. After seed maturation, all the phytometer plants were individually harvested and threshed. Seeds originating from a single phytometer mother plant are called seed family in the following text. Four of the eight replicate field plots per background line received fertiliser when the plants had reached phenological stages 11 and 39 (each time 3 g N m^{-2} were applied as "Ammonsalpeter 27.5", Lonza, Visp, Switzerland; see Zeller et al. 2010 for further details of field design).

The second part of experiment 1 took place from March–August in 2010. We planted offspring seeds of the eight phytometer lines from the field experiment 2008 back to the field site in Zurich-Reckenholz. Only phytometer plants that had flowered at the same time as the corresponding pollen-donating background plants and which produced at least four seeds were used. In total, 146 out of 265 seed families (4 blocks x 2 fertiliser treatments x 4 background lines x 8 phytometer lines) met these criteria. A minimum of 4 and a maximum of 16 seeds were planted from each seed family, resulting in a total of 1945 individual offspring. We hand-sowed the seeds in patches of four per seed family into ten 1 x 4 m plots. The patches were assigned to positions and plots in a completely randomised design. The positions within a plot formed a grid of three rows with a distance of 18 cm between patches between and within rows (60 seed patches per plot). The plots were arranged in a grid aligned along an x-axis leading from east to west and a y-axis leading from south to north. The plots were surrounded by additional buffer plants of variety Bobwhite to avoid edge effects on the test plants. Phosphorus and potassium fertiliser had been applied to the plots prior to the seed planting in autumn 2009 at a rate of 46 kg P2O5 ha^{-1} and 60 kg K2O ha^{-1}. The amount of mineralised nitrogen, determined at the end of February 2010 in the top 100 cm of the soil was 41.7 kg N ha^{-1}. Nitrogen fertiliser was additionally applied immediately after sowing (30 kg N ha^{-1}) and another 30 kg N ha^{-1} when plants had reached the phenological stage 39 (Zadoks et al. 1974). All plots were sprayed with the herbicide cocktail Concert SX (40% Thifensulfurone, 4% Metusulfurone-methyl; Stähler Suisse AG, Zofingen, Switzerland) on 18 May.

We determined the cross-pollination rate by dividing the number of mature offspring hybrids through the total number of mature plants per phytometer. Hybrids produced by cross-pollination of Bobwhite by Frisal or Casana differed visibly in their morphology from offspring produced by self-pollination or cross-pollination with other Bobwhite plants. They were taller and had a reduced awn length than the parental varieties and suffered from slight hybrid necrosis (O. Kalinina et al., unpublished data), which can occur when unrelated wheat varieties are crossed (Hermsen 1963).

All hybrids and 65 randomly chosen putatively selfed offspring were tested for the presence or absence of the transgenes *Pm3b*, *chi* and *chi/glu* using Polymerase Chain Reaction (PCR) analysis. DNA was isolated from 200–300 mg of fresh leaf tissue by adapting the method of Stein et al. (2001). For the amplification of the *Pm3b* gene, we chose primer sequences fitting the ubiquitin promoter (5'-ATCTCTGTCGCTGCCTCTGG-3' and 5'-TGTGCGCTCCGAACAACACG-3'; Sigma-Aldrich GmbH, Buchs, Switzerland). The *chi/glu* transgenes were detected by amplification of parts of the *bar* gene in the MAGUCM plasmid ('5-TCAACCACTACATCGAGACA-3' and '5-AGTCCAGCTGCCAGAAAC-3'; Sigma-Aldrich GmbH, Buchs, Switzerland). The amplified DNA was separated and visualized performing gel electrophoresis. In total, 97.5% of the hybrids and the putatively selfed offspring were identified correctly, based on the presence/absence tests of *Pm3b*, *chi*, and *chi/glu* transgenes (data not shown). We conclude therefore that the method of hybrid detection by visual phenotyping was appropriate.

To quantify the morphological differences between hybrids (Table S1) and parental varieties we measured the height and awn length of all hybrids and the 65 randomly chosen putatively selfed offspring. Plant height was defined as the highest point of the plant measured from the soil at the end of the growing season. The awn length was determined by measuring the distance between the top of a spike and the tip of the awn. Furthermore we assessed the number of spikes, the number of seeds and the seed yield, which is equivalent to the total seed mass.

After the seedlings emerged, we observed damage by larvae of the click-beetle *Agriotes lineatus*. Click-beetle larvae, called wireworms, are known worldwide as pests of a range of agricultural crops including wheat (Parker and Howard 2001; Vernon et al. 2009). The wireworms damaged or killed around 30% of the planted seeds. We treated all plots with the insecticide Dyfonate 5G (Lesco Inc., Cleveland, USA) and replaced the damaged plants with seedlings of the variety Bobwhite to keep the conditions for the remaining plants comparable. To test the impact of the damage we counted the number of dead vs. surviving plants for further analysis.

Experiment 2: Short-distance gene flow between adjacent subplots

The second field experiment also took place at the research station in Zurich-Reckenholz and lasted from March–August 2009. Three GM lines *Pm3b*#1, *Pm3b*#2 and A9 *chi* and their corresponding non-GM lines S3b#1, S3b#2 and Frisal were grown in three separate 7 x 1 m cross-pollination plots (Figure S1). Each plot consisted of one subplot (1 x 1 m) in the centre with GM plants as pollen donors and four subplots (0.5 x 1 m) on two opposing sides with the corresponding non-GM plants as pollen recipients. The opposing sides were in eastern or western direction of the pollen source because the prevailing winds at the field site blow from the west (Figure S1). The distances between central subplot and side subplots were 0–0.5, 0.5–1, 1–1.5 and 2–2.5 m (a subplot also occurred between 1.5–2 m but was not harvested). As there were four replicate blocks, three line combinations and eight subplots with pollen recipients (distance subplots), the sample size was 32 for each tested line and 96 in total. The distance subplots were sown with an Oyjord plot drill system (Wintersteiger AG, Switzerland) and the central plots with the GM pollen source by hand. Seeding density was 400 seeds/m^2 and there were six rows spaced 18 cm apart. The cross-pollination plots were at least 2 m apart from each other and the intervals were filled with tall-growing Triticale plants acting as a pollen barrier to minimise cross-pollination between plots. Flowering periods of pollen donor and receptor subplots were similar in order to allow cross-pollination. Nitrogen fertiliser was applied one day before sowing (40 kg N ha^{-1}) on 25 March and again when the plants had reached the phenological stage 11 (30 kg N ha^{-1}). Phosphorus and potassium fertiliser were applied at the phenological stages 11 and 37 (Zadoks et al. 1974) at a rate of 46 kg P2O5 ha^{-1} and 60 kg K2O ha^{-1}. The plots were sprayed with the herbicide cocktail Concert SX (40% Thifensulfurone, 4% Metusulfurone-methyl; Stähler Suisse AG) and Starane super (120 g l^{-1} Bromoxynil, 120 g l^{-1} Ioxynil, 100 g l^{-1} Fluroxypyr-metilheptil-ester; Omya Agro AG, Safenwil, Switzerland) in the beginning of May. The plots were treated twice with the insecticide Karate Zeon (100g l^{-1} Lambda-Cyhalothrin; Syngenta Agro AG, Dielsdorf, Switzerland) against the wheat stem fly (Chlorops pumilionis Bjerk.) in the beginning of May and 2 weeks later.

To measure the cross-pollination rate in each distance subplot we used a novel population-level PCR analysis that detected the transgenes *Pm3b* and A9 *chi* in batches of seeds. A single-seed approach was not feasible due to the low expected cross-pollination rates. The optimal size of seed batches was determined in a pilot study with flour from seed batches of defined numbers of GM and non-GM seeds. PCR amplification of DNA extracted from flour of the different seed batches showed that a single GM seed could be detected reliably in 1:10, 1:50, 1:200 and 1:500 mixtures of GM:non-GM seeds. Potential outcrosses would be heterozygous and would therefore contain only 50% of the DNA of a homozygous GM seed. Taking this into account, we opted for seed batches of 100 seeds in our experiment 2 (Figure S2).

For the analysis of the cross-pollination rate, we collected 5 batches of 100 seeds per distance subplot and produced flour from each batch (TissueLyser, Qiagen Instruments AG, Hilden, Germany). To avoid DNA contamination between batches, the jars used for the milling were sprayed with DNA-ExitusPlusTM (AppliChem GmbH, Darmstadt, Germany) and incubated at 60 °C for 10 min to increase the degradation rate of DNA (Esser et al. 2006).

DNA was extracted from 20 mg flour per sample adapting the method of (Kang et al. 1998). To test the DNA extracts for transgene-presence we used the same PCR protocol as described above. If a sample tested positive, DNA extraction and PCR were repeated. Positive samples were therefore based on at least two independent DNA extractions and PCR reactions (Figure S3).

Data analysis

The influence of background and phytometer lines on cross-pollination within the plot, measured as the probability of an individual offspring plant to be a hybrid rather than a putatively selfed offspring (experiment 1) was analysed using generalized linear models (GLMs) with logit link function and binomial error distribution (McCullagh and Nelder 1989). To account for potential overdispersion, experimental factors were tested with approximate F-tests derived from analysis of deviance tables (Crawley 2007; see Table S2). Experimental factors were background line, phytometer line, phytometer nutrient environment and phytometer individual (seed family). Plants that did not germinate or died due to pest infestation were excluded from further analysis.

Differences in height, awn length, seed number and seed yield of hybrid and putatively selfed offspring in experiment 1 were analysed with ordinary GLMs with identity link function and normal errors. Because these traits were not influenced by the experimental design in which the seeds were produced in 2008; we did not fit the corresponding experimental factors for block and fertiliser application. Two factorial models were fitted to analyse either the line variation within hybrid and non-hybrid plants or general effects of the lines and their hybridisation (Tables S3 and S4). Effects of background (pollen donor or father) and phytometer (pollen recipient or mother) line were analysed separately. We examined residual plots to identify outliers and to check if the assumptions of normality and homoscedasticity were fulfilled. One unusually high-yielding plant was identified as an outlier an excluded from the analysis of seed yield.

The seedling mortality of offspring was also analysed using a GLM with logit link function and binomial error distribution. In this case, experimental factors describing the growing conditions of parental plants in the field in 2008 (block and fertiliser application) were also included in the statistical model. Furthermore, because mortality was not evenly distributed in the field in 2010 we also fitted two position covariates in the form of x- and y-coordinates of the planting grid (Table S5).

Data from experiment 2, the short-distance gene-flow experiment, were analysed using GLMs with logit-link function and binomial error distribution (Table S6). The dependent variable was the probability to find a transgene in a batch of 100 seeds. In one model, the experimental factor distance was decomposed into a contrast log(distance) and residual variation between distance classes because cross-pollination rates are likely to decrease logarithmically with increasing distance to the pollen source (Albrecht et al. 2009). To investigate differences between very short (0–0.5 m) and short-distance (0.5–2.5 m) gene flow, we split the dataset and analysed both subsets separately. The highest possible estimate of cross-pollination rate was calculated by dividing the observed amount of positive batches by the total amount of batches. This makes the highly unlikely assumption that in all positive batches all 100 seeds result from cross-pollination. The lowest possible estimate of cross-pollination rate was calculated by dividing the positive samples by the total amount of samples multiplied by 100. This makes the assumption that in all positive batches only 1 seed out of 100 is the result of cross-pollination. Following the maximum likelihood estimation for binomial data, we calculated the values most likely to have produced the observed results (Fisher 1922). The estimate for the probability is: $p = 1 - ((n-z)/n)^{(1/J)}$, with n being the total amount of batches, while z represents the positive batches and J the number of seeds per batch, i.e. 100. All statistical analyses were performed with the statistical software R 2.9.2 (Team 2010). The critical significance level was 0.05 for all analyses.

Results

Experiment 1: gene flow within plots

40 out of 1192 mature plants could be identified visually as hybrids (Table S1) indicating that 3.36% of all planted seeds had received pollen from foreign wheat varieties (Background). Overall, 14.4% of all mother plants produced at least one hybrid seed and 19.6% of all seeds of such plants were identified as hybrids. 21 hybrids were crosses between two GM lines leading to natural pyramiding of *Pm3b* and *chi* or *chi/glu* transgenes.

The identity of the mother line influenced the hybridisation rate ($P < 0.001$ for differences among mothers): 7.25% of all *Pm3b* seeds were hybridised, which is 6.2 times as many as for corresponding control phytometers ($P < 0.001$ for difference GM/controls). *Pm3b*#2 had fewer hybrids than the other *Pm3b* ($P = 0.018$ for difference *Pm3b*#2/residual *Pm3b*, Figure 1). Also, the background, i.e. the father side influenced the hybridisation rate: among the Frisal fathers, A9 *chi* pollinated more plants than did A13 *chi/glu* ($P < 0.001$ for difference A9/A13, Figure 2). Finally, there were some significant interactions between mother and father lines (Table S2). The Frisal control fathers pollinated more control mothers than did the Frisal GM fathers, which in turn pollinated preferably GM (=*Pm3b*) mothers ($P = 0.03$ for interaction *Pm3b* vs. control x Frisal GM vs. Frisal control).

Besides the hybridisation rate, we measured awn length, plant height, spike number, seed number and seed yield of all hybrids and some randomly chosen non-hybridized plants. We found that awn length was higher towards the south and east of the experimental area ($P < 0.001$ for x-axis and $P = 0.001$ for y-axis, Table S3). This could have been due to variation in soil properties or the occurrence of soil pathogens. All other traits were not influenced by the geographical position of plants. *Pm3b*#2 and #4 had longer awns, fewer seeds and lower yield than *Pm3b*#1 and #3 ($P = 0.004$ for difference in awn length, $P = 0.02$ for difference in seed number and $P < 0.001$ for difference in yield for *Pm3b*#2 and #4/*Pm3b*#1 and #3). Hybrids with Bobwhite mothers and Frisal fathers were 6.3% (5.19 ± 0.49 cm) taller, had 37.3% (1.4 ± 0.14 cm) shorter awns and 26.9% (1.4 ± 0.16 g) lower seed yields than Bobwhite mother plants. These trait differences were significant even if all line effects were fitted before the term hybridisation (Table S4).

Hybrids with Casana fathers had shorter awns than hybrids with Frisal fathers (P = 0.013 for difference Casana father/Frisal father). Within fathers of the variety Frisal, plants pollinated by Frisal GM fathers (A9 *chi*, A13 *chi/glu*) had shorter awns than when pollinated by Frisal control fathers (P = 0.02 for difference Frisal A9 and A13 father/Frisal control father).

A total of 1192 out of 1945 plants (61.3%) reached maturity. The position of each within the experimental field (x- and y-axis and interaction) affected the mortality significantly. Plants in the northwest part of the experimental area died more often than plants in the southeast. This correlated with the occurrence of wireworms, which were observed in great numbers in the northwestern part. It is therefore likely that wireworms were responsible for the observed mortality. Seeds that came from mothers that were grown on fertilised plots died (or were eaten) 2.4 times as often as seeds from mothers that were grown on unfertilised plots (P < 0.001 for difference fertilised/not fertilised), suggesting that there were environmental maternal carry-over effects from mothers to offspring. Seedlings with a *Pm3b* transgene died 1.3 times as often as control plants (P = 0.005 for difference *Pm3b*/control plants, Figure 3). The mortality of *Pm3b*#2 plants (58%) was 1.4 times as high as that of other *Pm3b* plants (P < 0.001 for difference *Pm3b*#2/*Pm3b*#1, 3 and 4).

Experiment 2: Short-distance gene flow between adjacent subplots

Upper and lower boundaries of the estimated cross-pollination rates are shown in Figure 4A. The upper boundary shows cross-pollination rates assuming that all seeds of a 100-seed sample were genetically modified, if a single seed was tested positive, whereas the lower boundary assumes that only 1 seed in a 100-seed sample was positive. Using the log(distance) model, we found higher cross-pollination probabilities in the west than in the east (P = 0.048 for difference west/east, Table S6). Furthermore, Frisal A9 *chi* plants were more likely to outcross than Bobwhite plants (P = 0.02 for difference Bobwhite/Frisal A9 *chi*). We found no significant differences between the two *Pm3b* lines if we combined the data of all distances. However, if we analysed the subplots closest to the pollen source (0–0.5 m) separately, *Pm3b*#1 was more likely to outcross than *Pm3b*#2 (P = 0.05 for difference *Pm3b*#1/*Pm3b*#2). Neither varieties, lines nor wind directions differed significantly in subplots further away from the pollen source (0.5–2.5m).

The actual cross-pollination rates lie between the upper and the lower boundary estimates. We calculated the most likely cross-pollination rate for each distance subplot using a maximum likelihood method (Figure 4B). We found that the overall cross-pollination rate was 0.8% in the west and 0.5% in the east if measured at a distance of 0–0.5 m from the pollen source. Cross-pollination rates decreased sharply (logarithmically) with increasing distance to the pollen source. However, our methods were accurate enough to detect cross-pollination events in 2.5 m distance to the pollen source. The detected rates of 0.05% in the west and 0.02% in the east would be enough to meet the seed purity levels set by the European Union (Beckie and Hall 2008).

Discussion

Increased gene flow of Pm3b wheat lines within the field

Large differences among wheat cultivars concerning pollen-mediated gene flow have been reported before and were often attributed to dissimilarities in male fertility and morphological traits (Waines and Hegde 2003; Matus-Cádiz et al. 2007). However, we found no published reports studying the influence of resistance genes on gene flow. Our results show remarkable differences in gene flow between the Bobwhite GM (*Pm3b*) and control lines and among the different *Pm3b* lines. This confirms the results of a previous study in which depending on the insertion event, a particular transgene had a large influence on the entire phenotype (Zeller et al. 2010). One reason could be resistance costs due to transgene expression (Strauss et al. 2002; Stowe and Marquis 2010; Bergelson and Purrington 1996), which may have weakened the general fitness of the plants and possibly have caused reduced fertility. Lines *Pm3b*#2 and #4 had strongly increased levels of ergot infection, probably due to altered spike morphology during flowering (Zeller et al. 2010). A possible explanation for the altered spike morphology might have been reduced pollen quality or quantity. As a reaction, the plants could have made their florets more accessible to foreign pollen, as other studies with male-sterile lines have shown (Waines and Hegde 2003). We therefore expected increased hybridisation rates for the lines *Pm3b*#2 and #4, yet surprisingly *Pm3b*#2 mother plants had a lower hybridisation rate than mother plants of all other *Pm3b* lines. A reason for the low hybridisation rate of *Pm3b*#2 might have been the low general performance of this line in the field which may also have affected female function (Zeller et al. 2010). Besides the capacity to receive foreign pollen, the ability to pollinate other plants seems to be important to gene flow as indicated by the differences in pollination rates between father plants from different lines.

We found that on average 3.36% of all tested seeds had resulted from hybridisation with neighbouring plants. However, this cross-pollination rate varied among the eight wheat lines tested from 0.5% and 8.5%. These rate measurements are critical to answer questions concerning the EU 0.9% threshold for GM seeds in the harvest (Graziano et al. 2007). A study with maize (*Zea mays* L) using a colour marker showed an increased percentage of marked seeds at harvest compared to sowing (Dietiker et al. 2011). The contamination percentage at sowing was 1% and on average 2.8 times as high at harvest. The authors therefore concluded that contamination at sowing should be as low as 0.2–0.5% to guarantee the EU 0.9% threshold at harvest. In other words, in the case of maize a seed purity of 0.9% at sowing would not be sufficient to ensure the threshold. However, in the mainly selfing crop wheat, the increase in percentage GM seeds from sowing to harvest would be much smaller even under worst-case scenarios: assuming a cross-pollination rate of 8.5% (the maximum found above) and an initial GM proportion of 0.9% the proportion at harvest would rise to 1.084% (seeds which are homo- or heterozygous for the transgene). Obviously, the occurrence of GM plants in wheat fields planted with male-sterile plants could increase the possibility of unintended hybridisation.

Phenotypic effects of heterozygocity

In wheat, heterozygotic effects have been documented before (Borghi and Perenzin 1994; Perenzin et al. 1998). Hybrids were found to develop faster and grow taller. Artificial crosses of Frisal (tall, short awns) and Bobwhite (short, long awns) resulted in tall hybrids with relatively short awns (O. Kalinina et al., unpublished data). In view of these findings, we expected hybrids between Bobwhite mother plants and Frisal fathers to differ significantly in height and awn length from non-hybridised control plants. In our case, hybrids had shorter awns and grew taller than non-hybridised Bobwhite mothers, but they also a significantly lower seed yield. The latter is an uncommon pattern found in line-hybrids within plant species (Borghi and Perenzin 1994; Perenzin et al. 1998). However, our hybrids also suffered from necrotic leaf tips, a phenomenon known for hybrids between unrelated lines in wheat (Hermsen 1963). It is likely that the genetic differences between the Mexican variety Bobwhite and the Swiss varieties Frisal and Casana were too large for heterosis effects such as hybrid vigor to show up.

From previous experiments we expected to find differences in seed yield between Bobwhite GM and control lines (Zeller et al. 2010) but we could not find such differences in the present experiment 1. We could, however, find similar differences as in the previous experiment among the different Bobwhite GM lines: *Pm3b*#2 and *Pm3b*#4 had lower seed yield than the lines *Pm3b*#1 and *Pm3b*#3. Very high transgene expression levels in the first two lines could have led to the reduced seed yield. Hybrids with Frisal GM fathers had shorter awns than hybrids with Frisal control fathers. This may indicate an unintended influence of the inserted *chi* and *chi/glu* transgenes on the offspring phenotype.

Finally we would like to discuss the method used to study cross-pollination within a field. Hybrids could be identified easily due to phenotypic effects. The reliability of hybrid detection was confirmed performing PCR. This approach seems to be a good alternative to time-consuming DNA sequencing of entire plant populations. It can be used for all wheat varieties that show strong heterozygous effects when crossed with other varieties.

Maternal effects and transgene influence on seedling mortality

We found that the mortality of seeds produced in plots with added fertiliser was higher than that of seeds from other plots. Nitrogen addition tends to increase the seed size and nitrogen content, which is proportional to the protein content (S. Zeller, unpublished data). Such maternal effects (see e.g. Steinger et al. 2000) are usually known to increase the fitness of individual seeds. However, in our experiment the driving force behind the seedling mortality seemed to be wireworms that occurred in great numbers within the experimental plots. The most evident explanation is that the improved nutrient content made these seedlings more attractive to wireworms.

Besides these maternal effects, GM lines of Bobwhite had higher seedling mortality than control lines of Bobwhite. Previous studies, however, found no difference in seed germination and seedling mortality among different lines (O. Kalinina, unpublished data). The differences observed in the present study may again be explained by the wireworm infestation: it is conceivable that *Pm3b* lines paid an ecological cost of resistance due to transgene expression (Strauss et al. 2002). Costs of resistance, which were discovered in several other studies (Bergelson and Purrington 1996; Stowe and Marquis 2010), could have reduced a potential natural defence against wireworms in Bobwhite GM plants or compromised their ability to survive the attack. Wireworms feed on live vegetable material (seeds, stems, roots) to survive and grow (Furlan 1996). GM lines may have been preferable forage because of their slower development, smaller seed size or softer skin and/or seed material. Since the increased mortality of the *Pm3b* lines seemed to depend on the occurrence of the wireworms, it can be seen as a transgene x environment interaction.

Gene flow in wheat: short and random

The estimated gene flow over distance matches the results of prior observations in which the average cross-pollination rate was about 1% in close proximity and decreased rapidly with distance from the pollen source (Gustafson et al. 2005). When planning our experiment, we expected to find stronger cross-pollination toward the east than toward the west, due to prevailing winds at the study site. As expected, winds mostly blew from west or northwest during flowering (data not shown). Surprisingly, however, we observed higher cross-pollination rates in the western subplots. Data from a nearby weather station showed that there were a few hours of light easterlies or north-easterlies while 23% of the mother plants were flowering. It might be that cross-pollination occurred mainly during those hours, which then led to a higher cross-pollination in the western subplots. Gene flow also occurred mostly in the opposite direction of prevailing winds in a study by Gatford et al. (2006). We conclude, therefore, that not only prevailing winds are important for cross-pollination, but the winds at the exact time of flowering. Hence, as the time of flowering in wheat is usually short (De Vries 1971), cross-pollination can occur in all directions. This should be considered when planning cross-pollination experiments and determining isolation distances.

We could detect significant differences in gene flow between the varieties Bobwhite and Frisal over a distance of < 0.5 m. When comparing the varieties from the subplots which were at least 0.5 m away from the pollen source, no significant differences could be detected anymore. *Pm3b*#1 only outcrossed significantly more than *Pm3b*#2 up to a distance of 0.5 m from the pollen source. We conclude therefore that the differences between varieties and lines are mainly present over short distances between pollen donor and recipient.

Our results indicate that pooling can be an appropriate method to gain information on an entire population. Taking population samples of 100 pooled seeds turned out to be the optimal size to estimate rates of cross-pollination over short distances using a maximum-likelihood method. Pooled measures over larger distances or individual measures even over the shortest distance would have led to (too) many negative counts.

Conclusions

Our results show that inserting a single transgene into a wheat genome can significantly affect the ecological behaviour of the resulting plant lines. We found that Bobwhite mother plants with a *Pm3b* transgene were more likely to hybridize with other wheat varieties than were non-GM Bobwhite mother plants. Furthermore, father line A9, which harboured a *chi* transgene, produced more offspring than non-GM Frisal father plants. We could demonstrate that hybrids with two or even three transgenes can occur if different GM plants are planted in close proximity. Such plants could further complicate environmental risk assessments. Besides differences in cross-pollination rates, we found that Bobwhite plants with *Pm3b* transgenes seemed to suffer ecological costs of resistance. All of these lines, but especially the one with the highest transgene expression, *Pm3b*#2, had increased seedlings mortality if attacked by wireworms.

Because cross-pollination rates varied strongly between varieties and among GM lines it may be difficult to develop universal models for pollen-mediated gene flow in wheat. A case-by-case approach might be more promising (Andow and Zwahlen 2006). The gene-flow rates which we measured in our experiment 2 indicate that gene flow in wheat mainly occurs over short distances. However within the field, 14.4% of all plants received pollen from and 3.4% of the offspring seeds were sired by neighbouring plants. Each homozygous GM plant is likely to outcross with several neighbours which will result in plants heterozygous for the transgene. The proportion of GM plants within a population is therefore likely to increase. If we take a cross-pollination rate of 3.4% and assume an initial GM contamination of 0.9%, 0.931% of all offspring seeds would contain at least one copy of the transgene. If all plants would have been cross-pollinated, this rate would increase to 1.79% in one generation only. We conclude that the determination of cross-pollination rates within the field might be more important than cross-pollination over a distance in order to define appropriate threshold limits necessary to allow coexistence of GM and conventional farming systems.

Acknowledgements

We thank Y. Hautier, X. Li and M. Zach for discussions and comments, S. Geinitz for mathematical advice, the national research station Agroscope Reckenholz-Tänikon ART for setting up the field experiments, S. Nägeli for volunteering and numerous helpers in the field for assistance, G. Herren, S. Brunner, C. Diaz Quijano and R. Husi for the assistance in the laboratory and Meteo Schweiz for providing climatic data. This project was supported by the Swiss National Science Foundation and is a part of the wheat-cluster.ch, a sub-unit of the national research programme NRP 59 (SNF 405940-115607).

References

Albrecht, M., Duelli, P., Obrist, M. K., Kleijn, D. & Schmid, B. (2009) Effective long-distance pollen dispersal in *Centaurea jacea*. *PloS ONE,* **4,** e6751.

Andow, D. A. & Zwahlen, C. (2006) Assessing environmental risks of transgenic plants. *Ecology Letters,* **9,** 196–214.

Beckie, H. J. & Hall, L. M. (2008) Simple to complex: modelling crop pollen-mediated gene flow. *Plant Science,* **175,** 615–628.

Bergelson, J. & Purrington, C. B. (1996) Surveying patterns in the cost of resistance in plants. *American Naturalist,* **148,** 536–558.

Bieri, S., Potrykus, I. & Futterer, J. (2003) Effects of combined expression of antifungal barley seed proteins in transgenic wheat on powdery mildew infection. *Molecular Breeding,* **11,** 37–48.

Bliffeld, M., Mundy, J., Potrykus, I. & Futterer, J. (1999) Genetic engineering of wheat for increased resistance to powdery mildew disease. *Theoretical and Applied Genetics,* **98,** 1079–1086.

Borghi, B. & Perenzin, M. (1994) Diallel analysis to predict heterosis and combining ability for grain-yield, yield components and bread-making quality in bread wheat (*Triticum aestivum*). *Theoretical and Applied Genetics,* **89,** 975–981.

Cellini, F., Chesson, A., Colquhoun, I., Constable, A., Davies, H. V., Engel, K. H., Gatehouse, A. M., Karenlampi, S., Kok, E. J., Leguay, J. J., Lehesranta, S., Noteborn, H. P., Pedersen, J. & Smith, M. (2004) Unintended effects and their detection in genetically modified crops. *Food and Chemical Toxicology,* **42,** 1089–1125.

Christensen, A. H. & Quail, P. H. (1996) Ubiquitin promoter-based vectors for high-level expression of selectable and/or screenable marker genes in monocotyledonous plants. *Transgenic Research,* **5,** 213–218.

Clements, F. E. & Goldsmith, G. W. (1924) *The phytometer method in ecology: the plant and community as instruments.* Carnegie Institution of Washington, Washington, USA.

Conner, A. J., Glare, T. R. & Nap, J. P. (2003) The release of genetically modified crops into the environment. Part II. Overview of ecological risk assessment. *Plant Journal,* **33,** 19–46.

Crawley, M. J. (2007) *The R book.* John Wiley & Sons, Chichester, UK.

Daniell, H. (2002) Molecular strategies for gene containment in transgenic crops. *Nature Biotechnology,* **20,** 581–586.

De Vries, A. P. (1971) Flowering biology of wheat, particularly in view of hybrid seed production - review. *Euphytica,* **20,** 152–170.

De Vries, A. P. (1974) Some aspects of cross-pollination in wheat (*Triticum aestivum* L.). 4. Seed set on male sterile plants as influenced by distance from the pollen source, pollinator: Male sterile ratio and width of the male sterile strip. *Euphytica,* **23,** 601–622.

Dietiker, D., Oehen, B., Ochsenbein, C., Westgate, M. E. & Stamp, P. (2011) Field simulation of transgenic seed admixture dispersion in maize with a blue kernel colour marker. *Crop Science,* **51,** doi: 10.2135/cropsci2010.06.0311.

Eastham, K. & Sweet, J. (2002) *Genetically modified organisms (GMOs): The significance of gene flow through pollen transfer.* European Environment Agency, Copenhagen, Denmark.

Esser, K., H. Marx, W. H. & Lisowsky, T. (2006) DNA decontamination: novel DNA-ExitusPlus™ in comparison with conventional reagents. *BioTechniques,* **40,** 238–239.

Fisher, R. A. (1922) On the mathematical foundations of theoretical statistics. *Philosophical Transactions of the Royal Society of London,* **222,** 309–368.

Furlan, L. (1996) The biology of *Agriotes ustulatus* Schaller (*Col, Elateridae*) .I. Adults and oviposition. *Journal of Applied Entomology,* **120,** 269–274.

Gatford, K., Basri, Z., Edlington, J., Lloyd, J., Qureshi, J., Brettell, R. & Fincher, G. (2006) Gene flow from transgenic wheat and barley under field conditions. *Euphytica,* **151,** 383–391.

Graziano, C. M., Bartlett, M. & Perrings, C. (2007) Landscape gene flow, coexistence and threshold effect : The case of genetically modified herbicide tolerant oilseed rape (*Brassica napus*). *Ecological Modelling,* **205,** 169–180.

Griffin, W. B. (1987) Outcrossing in New Zealand wheats measured by occurrence of purple grain. *New Zealand Journal of Agricultural Research,* **30,** 287–290.

Gustafson, D. I., Horak, M. J., Rempel, C. B., Metz, S. G., Gigax, D. R. & Hucl, P. (2005) An empirical model for pollen-mediated gene flow in wheat. *Crop Science,* **45,** 1286–1294.

Hedrick, P. W. (2004) *Genetics of populations.* Jones and Bartlett Learning, Sudbury, USA.

Hermsen, J. G. T. (1963) Hybrid necrosis as a problem for wheat breeder. *Euphytica,* **12,** 1–16.

Jørgensen, R. B. & Andersen, B. (1994) Spontaneous hybridization between oilseed rape (*Brassica napus*) and weedy *Brassica campestris* (*Brassicaceae*) - a rise of growing genetically-modified oilseed rape. *American Journal of Botany,* **81,** 1620–1626.

Kang, H. W., Cho, Y. G., Yoon, U. H. & Eun, M. Y. (1998) A rapid DNA extraction method for RFLP and PCR analysis from a single dry seed. *Plant Molecular Biology Reporter,* **16,** 1–9.

Leah, R., Tommerup, H., Svendsen, I. & Mundy, J. (1991) Biochemical and molecular characterization of three barley seed proteins with antifungal properties. *Journal of Biological Chemistry,* **266,** 1564–73.

Levin, D. & Kerster, H. (1974) Gene flow in seed plants. *Evolutionary Biology,* **7,** 139–220.

Martin, T. J. (1990) Outcrossing in twelve hard red winter wheat cultivars. *Crop Science,* **30,** 59–62.

Matus-Cádiz, M. A., Hucl, P. & Dupuis, B. (2007) Pollen-mediated gene flow in wheat at the commercial scale. *Crop Science,* **47,** 573–579.

McCullagh, P. & Nelder, J. A. (1989) *Generalized linear models (CRC Monographs on Statistics & Applied Probability).* Chapman and Hall, London.

Mercer, K. L. & Wainwright, J. D. (2008) Gene flow from transgenic maize to landraces in Mexico: An analysis. *Agriculture, Ecosystems & Environment,* **123,** 109–115.

Parker, W. E. & Howard, J. J. (2001) The biology and management of wireworms (*Agriotes* spp.) on potato with particular reference to the U.K. *Agricultural and Forest Entomology,* **3,** 85–98.

Perenzin, M., Corbellini, M., Accerbi, M., Vaccino, P. & Borghi, B. (1998) Bread wheat: F1 hybrid performance and parental diversity estimates using molecular markers *Euphytica,* **100,** 273–279.

Piñeyro-Nelson, A., Van Heerwaarden, J., Perales, H. R., Serratos-Hernandez, J. A., Rangel, A., Hufford, M. B., Gepts, P., Garay-Arroyo, A., Rivera-Bustamante, R. & Alvarez-Buylla, E. R. (2009) Transgenes in Mexican maize: molecular evidence and methodological considerations for GMO detection in landrace populations. *Molecular Ecology,* **18,** 750–761.

Pla, M., Paz, J.-L., Peñas, G., García, N., Palaudelmàs, M., Esteve, T., Messeguer, J. & Melé, E. (2006) Assessment of real-time PCR based methods for quantification of pollen-mediated gene flow from gm to conventional maize in a field study. *Transgenic Research,* **15,** 219–228.

Reed, J., Privalle, L., Powell, M. L., Meghji, M., Dawson, J., Dunder, E., Suttie, J., Wenck, A., Launis, K., Kramer, C., Chang, Y. F., Hansen, G. & Wright, M. (2001) Phosphomannose isomerase: An efficient selectable marker for plant transformation. *In Vitro Cellular & Developmental Biology-Plant,* **37,** 127–132.

Rieger, M. A., Lamond, M., Preston, C., Powles, S. B. & Roush, R. T. (2002) Pollen-mediated movement of herbicide resistance between commercial canola fields. *Science,* **296,** 2386–2388.

Snow, A. A., Andow, D. A., Gepts, P., Hallerman, E. M., Power, A., Tiedje, J. M. & Wolfenbarger, L. L. (2005) Genetically engineered organisms and the environment: Current status and recommendations. *Ecological Applications,* **15,** 377–404.

Stein, N., Herren, G. & Keller, B. (2001) A new DNA extraction method for high-throughput marker analysis in a large-genome species such as *Triticum aestivum*. *Plant Breeding,* **120,** 354–356.

Steinger, T., Gall, R., Schmid, B. (2000) Maternal and direct effects of elevated CO2 on seed provisioning, germination and seedling growth in *Bromus erectus*. *Oecologia,* **123,** 475–480.

Stowe, K. A. & Marquis, R. J. (2010) Costs of defense: correlated responses to divergent selection for foliar glucosinolate content in *Brassica rapa*. *Evolutionary Ecology*, doi 10.1007/s10682-010-9443–9.

Strauss, S. Y., Rudgers, J. A., Lau, J. A. & Irwin, R. E. (2002) Direct and ecological costs of resistance to herbivory. *Trends in Ecology & Evolution,* **17,** 278–285.

Team, R. D. (2010) R: A language and environment for statistical computing. R Foundation for Statistical Computing, Vienna, Austria.

Vernon, R. S., Van herk, W. G., Clodius, M. & Harding, C. (2009) Wireworm management I: Stand protection versus wireworm mortality with wheat seed treatments. *Journal of Economic Entomology,* **102,** 2126–2136.

Waines, J. G. & Hegde, S. G. (2003) Intraspecific gene flow in bread wheat as affected by reproductive biology and pollination ecology of wheat flowers. *Crop Science,* **43,** 451–463.

Weber, W. E., Bringezu, T., Broer, I., Eder, J. & Holz, F. (2007) Coexistence between gm and non-gm maize crops – tested in 2004 at the field scale level (Erprobungsanbau 2004). *Journal of Agronomy and Crop Science,* **193,** 79–92.

Wolfenbarger, L. L. & Phifer, P. R. (2000) The ecological risks and benefits of genetically engineered plants. *Science,* **290,** 2088–2093.

Yahiaoui, N., Srichumpa, P., Dudler, R. & Keller, B. (2004) Genome analysis at different ploidy levels allows cloning of the powdery mildew resistance gene Pm3b from hexaploid wheat. *Plant Journal,* **37,** 528–538.

Zadoks, J. C., Chang, T. T. & Konzak, C. F. (1974) Decimal code for growth stages of cereals. *Weed Research,* **14,** 415–421.

Zeller, S. L., Kalinina, O., Brunner, S., Keller, B. & Schmid, B. (2010) Transgene × environment interactions in genetically modified wheat. *PloS ONE,* **5,** e11405.

Zhu, Q., Maher, E. A., Masoud, S., Dixon, R. A. & Lamb, C. J. (1994) Enhanced protection against fungal attack by constitutive coexpression of *chitinase* and *glucanase* genes in transgenic tobacco. *Nature Biotechnology,* **12,** 807–812.

Figure 6: Photograph taken by Simon Zeller

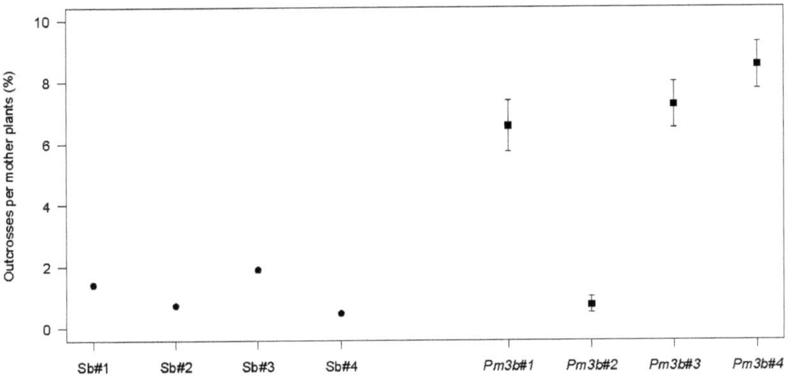

Figure 1: Cross-pollination rates of the eight pollen recipient lines (Bobwhite phytometer plants). Non-GM control lines (circles = S3b #1–4) had significantly lower cross-pollination rates than GM lines (squares = *Pm3b*#1–4). The GM line *Pm3b*#2 with highest transgene expression and lowest fertility had significantly lower cross-pollination rates than the other GM lines. Cross-pollination is defined as number of seeds derived from cross-pollination divided by number of all seeds x 100. Error bars represent ± 1 standard error (back-transformed from logit scale) and are sometimes hidden behind the symbols.

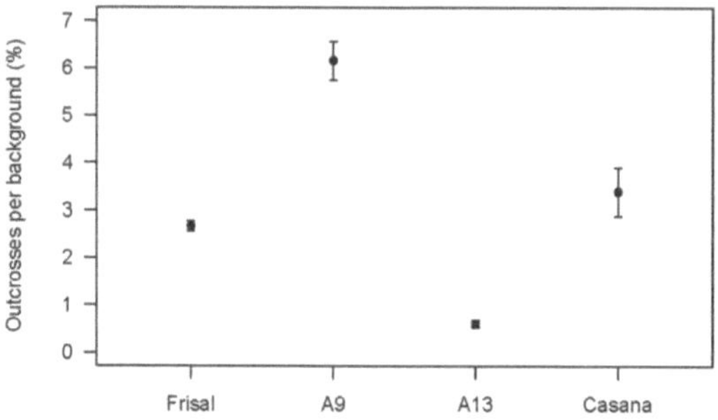

Figure 2: Cross-pollination rates of the four pollen-donor lines (background plants). Frisal and Casana are non-GM wheat varieties; A9 and A13 are GM lines based on the variety Frisal. The GM line A9 pollinated significantly more phytometer plants than did GM line A13. Cross-pollination (on pollen-receiving maternal plants) is defined as number of seeds derived from cross-pollination divided by number of all seeds x 100. Error bars represent ± 1 standard error (back-transformed from logit scale) and are sometimes hidden behind the symbols.

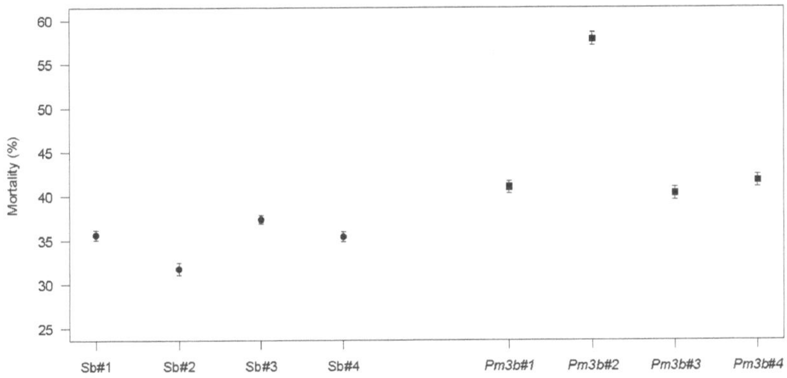

Figure 3: Mortality of seedlings produced by the eight pollen recipient lines (Bobwhite phytometer plants) by self- or cross-pollination. GM lines (squares = *Pm3b*#1–4) had higher mortality rates than corresponding non-GM control lines (circles = S3b #1–4) and *Pm3b*#2 had a higher mortality rate than *Pm3b*#1, 3 and 4. Error bars represent ± 1 standard error (back-transformed from logit scale).

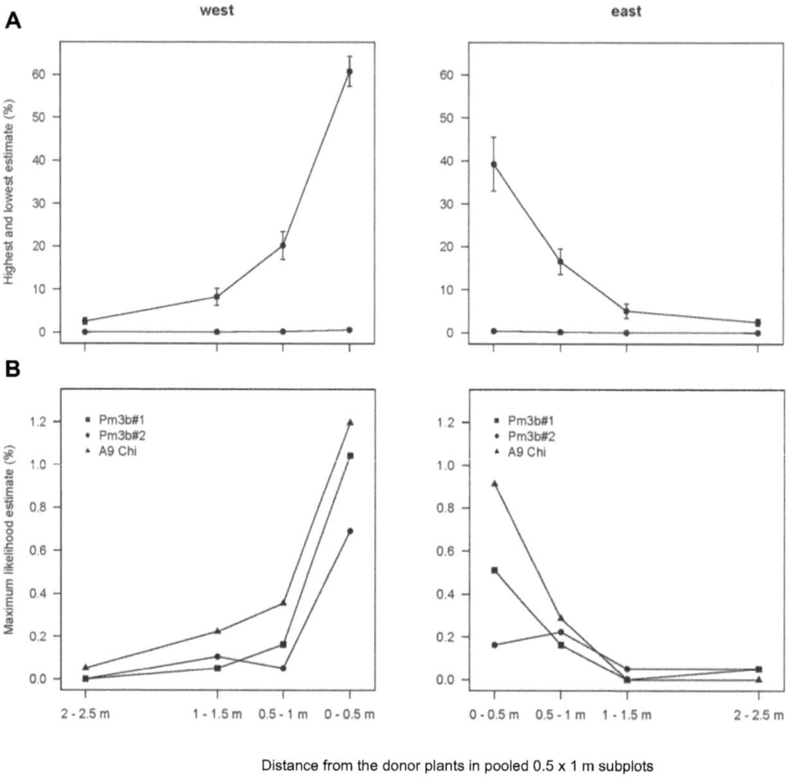

Figure 4: Cross-pollination of GM wheat over distance and in two wind directions. A: Upper and lower boundaries of cross-pollination rate estimates (mean±1 SE, back-transformed from logit scale) for western and eastern distance subplots. Data from all lines were pooled. B: Maximum likelihood estimate of cross-pollination rate for the western and eastern subplots for the lines *Pm3b*#1, *Pm3b*#2 and A9 *chi*. These estimates indicate cross-pollination rates between 1.2% and 0.16% in the closest and 0.05% and 0.0% in the farthest subplots.

Supplementary information

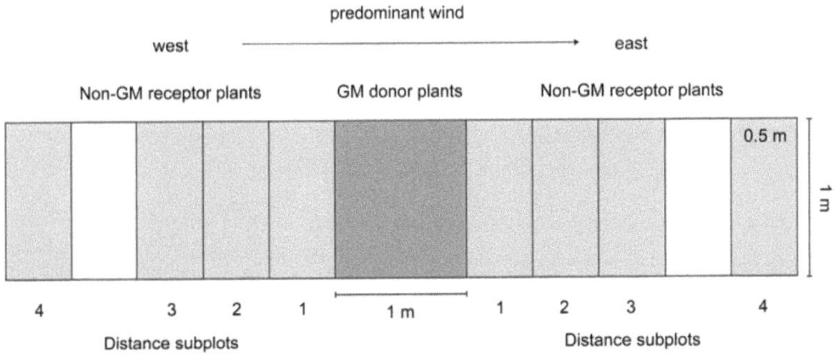

Figure S1: Schematic design of a cross-pollination plot. In the centre a 1 m^2 quadratic subplot of GM wheat was sown as a pollen source. In the eastern and western direction corresponding non-GM plants were sown as pollen recipients into distance subplots (0.5 x 1 m). The lightly shaded distance subplots were harvested after seed maturation.

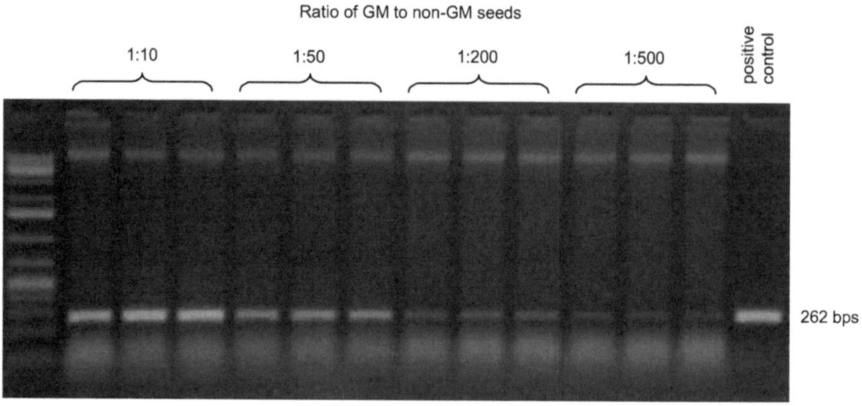

Figure S2: PCR analysis from flour of different seed mixtures containing 10%, 2%, 0.5% and 0.2% GM seeds. The positive bands show decreasing signal strength as the proportion of GM seed material decreases. Each analysis was replicated three times.

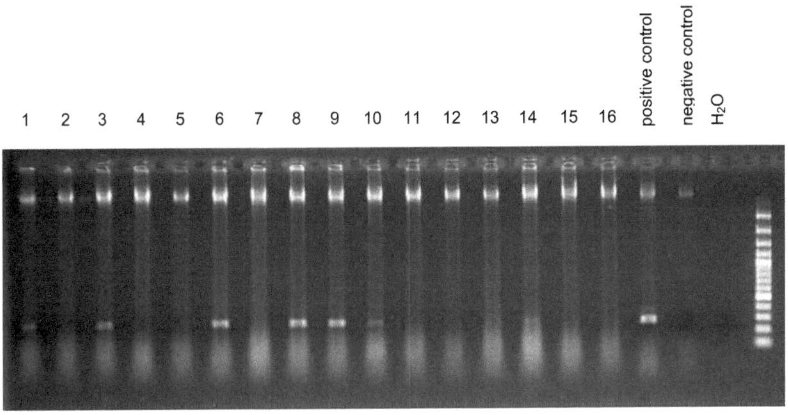

Figure S3: PCR analysis of wheat flower shows presence or absence of transgenes. Distinct white bands at the same height as the positive control indicate successful amplification of transgenic promoter regions (columns nr. 1, 3, 6, 8, 9, 10).

Table S1. Phenotypic data on hybrid plants. This table lists height, awn length, spike number, seed number and seed yield of all hybrid plants. The parents of each hybrid are indicated in the columns "mother" (pollen recipient line) and "background" (pollen donor line).

	Mother Line	Background Line	Plant height (cm)	Awn length (cm)	Spike number	Seed number	Seed yield (g)
GM lines	*Pm3b*#1	A13 *Chi/Glu*	86.7	2.2	4	84	2.53
	Pm3b#1	Frisal	89.7	7.4	5	152	3.64
	Pm3b#1	Casana	57.6	0.3	3	71	1.51
	Pm3b#1	Casana	86.6	0.6	3	119	3.66
	Pm3b#1	Casana	86.7	1.7	4	178	5.24
	Pm3b#1	Casana	87.1	0.8	6	238	6.98
	Pm3b#1	A13 *Chi/Glu*	92.1	1.9	3	77	2.77
	Pm3b#1	Casana	91.1	1.1	9	413	11.89
	Pm3b#1	Casana	88.7	0.6	7	342	10.56
	Pm3b#2	Frisal	87.0	3.6	4	137	3.34
	Pm3b#3	A9 *Chi*	89.5	3.5	3	86	1.93
	Pm3b#3	A9 *Chi*	94.2	3.1	5	156	3.78
	Pm3b#3	A9 *Chi*	96.8	3.3	5	212	4.83
	Pm3b#3	A9 *Chi*	90.2	3.2	3	113	2.74
	Pm3b#3	A9 *Chi*	82.5	1.1	2	53	1.47
	Pm3b#3	A9 *Chi*	82.0	1.8	4	127	2.68
	Pm3b#3	A9 *Chi*	86.3	1.3	3	141	3.99
	Pm3b#3	A9 *Chi*	93.3	1.5	7	235	3.75
	Pm3b#3	A9 *Chi*	93.7	0.8	4	185	5.98
	Pm3b#3	A9 *Chi*	101.0	2.2	5	208	5.99
	Pm3b#3	A9 *Chi*	88.2	1.3	15	572	16.35*
	Pm3b#4	A9 *Chi*	83.0	3.0	4	61	2.43
	Pm3b#4	A9 *Chi*	92.6	1.6	6	88	3.36
	Pm3b#4	A9 *Chi*	88.0	2.8	4	82	3.03
	Pm3b#4	Frisal	82.0	4.8	3	117	4.11
	Pm3b#4	Frisal	88.6	3.0	3	59	2.03
	Pm3b#4	A9 *Chi*	83.7	1.7	4	81	2.27
	Pm3b#4	A9 *Chi*	91.8	2.7	4	64	2.45
	Pm3b#4	A9 *Chi*	89.5	3.5	2	32	1.32
	Pm3b#4	A9 *Chi*	82.6	3.3	5	65	2.34
	Pm3b#4	A9 *Chi*	83.9	3.1	4	97	3.06
Control lines	Sb#1	A9 *Chi*	96.0	3.8	5	191	6.64
	Sb#1	Frisal	76.3	4.3	3	121	3.60
	Sb#1	Frisal	87.6	3.4	6	202	4.25
	Sb#2	A9 *Chi*	88.5	2.3	3	77	2.04
	Sb#3	Frisal	91.6	1.5	5	130	4.07
	Sb#3	Frisal	88.6	0.5	2	90	2.46
	Sb#3	A9 *Chi*	92.0	1.7	7	272	6.58
	Sb#3	Frisal	84.6	2.4	2	62	0.89
	Sb#4	Casana	92.5	2.0	4	93	2.75

Table S2. Factors influencing the rate of cross-pollination of GM and non GM-wheat in experiment 1. This analysis of deviance table shows the effects of block and fertiliser (conditions under which parental plants grew in the 2008 field experiment), pollen recipient line identity (one contrast between Bobwhite GM and non-GM lines: "*Pm3b* vs. control"; two contrasts within GM lines: "*Pm3b*#2 vs. residual *Pm3b* lines" and "within residual *Pm3b* lines"; residual differences between control lines: "Control"), pollen donor line identity (one contrast between Frisal and Casana: "Frisal vs. Casana"; one contrasts between Frisal GM and non-GM lines: "Frisal GM vs. Frisal control"; one contrast between GM lines: "Frisal A9 vs. Frisal A13") as well as their interaction on the rate of cross-pollination. Abbreviations: df = degree of freedom, % DV = % deviance change due to addition of terms to model, F pr. = error probability based on approximate F-ratios (ratios of mean deviance changes).

Source of variation	df	% DV	F pr.
Block	3	3.1	0.052
Fertiliser application	1	0.2	0.461
Pm3b vs. control	1	16.1	< 0.001
*Pm3b#*2 vs. residual *Pm3b* lines	1	2.3	0.018
Within residual *Pm3b* lines	2	0.5	0.566
Control	3	1.2	0.400
Frisal vs. Casana	1	0.1	0.566
Frisal GM vs. Frisal control	1	0.0	0.924
Frisal A9 vs. Frisal A13	1	9.6	< 0.001
Pm3b vs. control x Frisal vs. Casana	1	0.0	0.935
Pm3b vs. control x Frisal GM vs. Frisal control	1	1.9	0.030
Pm3b vs. control x Frisal A9 vs. Frisal A13	1	0.2	0.444
*Pm3b#*2 vs. residual *Pm3b* x Frisal vs. Casana	1	0.3	0.352
*Pm3b#*2 vs. residual *Pm3b* x Frisal GM vs. Frisal control	1	0.7	0.183
*Pm3b#*2 vs. residual *Pm3b* x Frisal A9 vs. Frisal A13	1	0.0	0.992
Control x Frisal vs. Casana	3	3.5	0.035
Control x Frisal GM vs. Frisal control	3	1.3	0.365
Control x Frisal A9 vs. Frisal A13	3	0.0	1.000
Within residual x Frisal vs. Casana	2	1.1	0.260
Within residual *Pm3b* x Frisal GM vs. Frisal control	2	7.0	< 0.001
Within residual *Pm3b* x Frisal A9 vs. Frisal A13	2	7.6	< 0.001
Residual	110	43.3	
Total	145	100	

Table S3. Factors affecting phenotypic traits of hybrid and non-hybrid wheat plants in experiment 1. This ANOVA table shows the effects of planting position (x- and y-coordinates), cross-pollination (of Bobwhite mother lines by Frisal or Casana pollen donors; "hybrids vs. non-hybrids) and pollen recipient line identity ("mother lines") as well as their interactions on plant height, awn length, seed number and seed yield. This model was used to test for the differences between mother lines after adjustment for hybrids vs. non-hybrids. Here the effects of mother lines are decomposed into one contrast between Bobwhite GM and non-GM lines ("*Pm3b* vs. control"), three contrasts within GM lines ("*Pm3b*#1 and #3 vs. *Pm3b*#2 and #4", "*Pm3b*#1 vs #3, "*Pm3b*#2 vs. #4") and residual differences between control lines ("Control"). Abbreviations: df = degree of freedom, % SS = % sum of squares explained by corresponding model terms, F pr. = error probability based on F-ratios.

		Plant height		Awn length			Seed number		Seed yield	
Source of variation	df	% SS	F pr.	% SS	F pr.	df	% SS	F pr.	% SS	F pr.
X-coordinate	1	0.6	0.410	16.2	< 0.001	1	0.4	0.5196	1.5	0.207
Y-coordinate	1	0.1	0.769	4.4	0.010	1	2.4	0.1058	1.5	0.207
X- x Y-coordinate	1	3.1	0.067	0.1	0.642	1	1.0	0.2997	0.5	0.450
Hybrids vs. non-hybrids	1	7.7	0.004	10.4	< 0.001	1	3.4	0.0540	5.3	0.018
Pm3b vs. control	1	1.1	0.273	2.1	0.075	1	0.1	0.6995	0.2	0.645
Pm3b#1 and #3 vs. *Pm3b*#2 and #4	1	0.4	0.503	2.8	0.040	1	5.1	0.0203	3.9	0.041
Pm3b#1 vs. #3	1	0.9	0.332	0.2	0.622	1	0.0	0.9940	0.7	0.392
Pm3b#2 vs. #4	1	2.8	0.080	1.6	0.114	1	1.6	0.1877	0.2	0.604
Control	3	3.5	0.278	1.4	0.549	3	4.5	0.1858	5.2	0.136
Hybrids vs. non-hybrids x *Pm3b* vs. control	1	1.3	0.235	0.1	0.651	1	1.0	0.2997	0.9	0.325
Hybrids vs. non-hybrids x *Pm3b*#1 and #3 vs. *Pm3b*#2 and #4	1	0.3	0.548	1.8	0.097	1	1.8	0.1643	0.6	0.402
Hybrids vs. non-hybrids x *Pm3b*#1 vs. #3	1	0.0	0.868	0.1	0.632	1	0.1	0.7325	0.3	0.549
Hybrids vs. non-hybrids x *Pm3b*#2 vs. #4	1	0.4	0.499	0.1	0.741	1	0.2	0.6538	0.2	0.609
Hybrids vs. non-hybrids x Control	3	1.0	0.781	3.6	0.143	3	1.4	0.6679	1.5	0.651
Residuals	86	76.9		55.2		85	77.0		77.4	
Total	104	100.0		100.0		103	100.0		100.0	

162

Table S4. Phenotypic effects of lines and their hybridisation in experiment 1. This ANOVA table shows the effect of planting position (x- and y-coordinates), pollen recipient line identity ("mother lines"), cross-pollination (of Bobwhite mother lines by Frisal or Casana pollen donors; "hybrids vs. non-hybrids) as well as their interactions on plant height, awn length, seed number and seed yield in experiment 1. This model was used to test hybrid effect after adjustment for differences between mother lines. Abbreviations: df = degree of freedom, % DV = % deviance change due to addition of terms to model, F pr. = error probability based on approximate F-ratios (ratios of mean deviance changes). Abbreviations: df = degree of freedom, % SS = % sum of squares explained by corresponding model terms, F pr. = error probability based on F-ratios.

		Plant height		Awn length			Seed number		Seed yield	
Source of variation	df	% SS	F pr.	% SS	F pr.	df	% SS	F pr.	% SS	F pr.
X-coordinate	1	0.6	0.410	16.2	< 0.001	1	0.4	0.520	1.5	0.208
Y-coordinate	1	0.1	0.769	4.4	0.010	1	2.4	0.106	1.5	0.207
X- x Y-coordinate	1	3.1	0.067	0.1	0.642	1	1.0	0.300	0.5	0.450
Mother lines	7	11.6	0.087	11.8	0.017	7	12.3	0.074	11.3	0.105
Hybrids vs. non-hybrids	1	4.8	0.023	6.6	0.002	1	2.5	0.104	4.3	0.034
Mother lines x hybrids vs. non-hybrids	7	3.0	0.846	5.7	0.272	7	4.5	0.667	3.6	0.782
Residual	86	76.9		55.2		85	77.0		77.4	
Total	104	100.0		100.0		103	100.0		100.0	

Table S5. Factors influencing offspring mortality in experiment 1. This analysis of deviance table shows the effects of the block, fertiliser (conditions under which parental plants grew in the 2008 field experiment), planting position (x- and y-coordinates) and pollen recipient line identity (one contrast between Bobwhite GM and non-GM lines: "*Pm3b* vs. control"; one contrast for differences between control lines: "Control"; two contrasts within GM lines: "*Pm3b*#2 vs. residual *Pm3b* lines" and "residual *Pm3b*") on offspring mortality caused by *Agriotes lineatus*. Abbreviations: df = degree of freedom, % DV = % deviance change due to addition of terms to model, F pr. = error probability based on approximate F-ratios (ratios of mean deviance changes).

Source of variation	df	% DV	F pr.
Block	3	1.7	0.281
Fertiliser application	1	3.1	0.008
X-coordinate	1	20.8	< 0.001
Y-coordinate	1	11.7	< 0.001
X- x Y-coordinate	1	2.2	0.026
Pm3b vs. control	1	3.5	0.005
Control	3	1.1	0.487
Pm3b#2 vs. residual *Pm3b*	1	5.3	< 0.001
Residual *Pm3b*	2	0.8	0.407
Residual	1929	50.0	
Total	1943	100.0	

Table S6. Factors influencing cross-pollination rates in experiment 2. This analysis of deviance table shows the effect of the wind direction (west vs. east), distance to pollen source (divided into log(distance) and residuals), plant variety and line identity as well as their interactions on the rate of cross-pollination. Abbreviations: df = degree of freedom, % DV = % deviance change due to addition of terms to model, F pr. = error probability based on approximate F-ratios (ratios of mean deviance changes).

Source of variation	df	% DV	F pr.
Block	3	1.1	0.429
West vs. east	1	1.5	0.048
log(distance)	1	41.4	< 0.001
Residual distance	1	0.0	0.913
Bobwhite vs. Frisal	1	3.7	0.002
Pm3b#1 vs. *Pm3b#2*	1	0.6	0.224
West vs. east x log(distance)	1	0.5	0.281
West vs. east x residual distance	1	0.0	0.768
log(distance) x Bobwhite vs. Frisal	1	0.2	0.531
Residual distance x Bobwhite vs. Frisal	1	0.1	0.640
log(distance) x *Pm3b#1* vs. *Pm3b#2*	1	0.9	0.127
Residual distance x *Pm3b#1* vs. *Pm3b#2*	1	0.1	0.656
Residual	81	50.0	
Total	95	100.0	

GENERAL DISCUSSION

Fig. 7: Discussion of sowing techniques during lunch break in spring 2010

The ecology of GM plants is the main topic of this thesis. A set of experimental spring wheat lines of the varieties Bobwhite (5 lines) and Frisal (2 lines) with additional fungal resistance genes were used as model organisms. Their performance was assessed comprehensively in multiple biotic and abiotic environments. We found that Bobwhite GM lines containing *Pm3b* transgenes were, as expected, more resistant to the fungus powdery mildew than their non-transgenic control lines (Chapter 1–4). In contrast to this, GM Frisal lines with *chitinase* or *chitinase* and *glucanase* transgenes showed no improved fungal resistance (Chapters 2 and 4). All tested GM wheat lines differed from their non-transgenic control lines in many traits, some of which were not directly linked to the additional fungal resistance. Unintended effects such as reduced yield, leaf necrosis and increased ergot infection were not visible in the glasshouse but only in the field experiments. Interestingly, even GM lines with identical transgenes differed strongly from each other; presumably due to differing transgene expression rates. Lines with the highest rates of *Pm3b* expression showed the strongest unintended phenotypic effects, costs of resistance, reduced competitive ability and increased seedlings mortality (Chapters 1–5). Costs of resistance were detected in all seven tested GM lines They were visible in absence but sometimes also in presence of fungal pathogens (Chapter 2). Mixtures of GM lines with different Pm3b alleles had higher yields than monocultures, presumably due to improved powdery mildew resistance. Finally, GM lines differed also in their ability to pollinate and receive pollen (Chapter 5). The detailed discussions of all results can be found at the end of each chapter. The aim of this section is to merge the main results of all five chapters and to discuss them in a wider, more applied context.

Improved powdery mildew resistance for lines with *Pm3* transgene

Wheat powdery mildew is a common wind-dispersed fungal pathogen that occurs in Africa, Asia, Australasia, Europe and throughout the Americas. Economically significant yield reductions occur in countries with high rainfall or irrigation with maritime and semi-continental climate (Bennett 1984). Plant breeders made large efforts to produce wheat varieties with powdery mildew resistance genes. We can see their success in our results. In our experiments we grew GM and non-GM plants of the Swiss variety Frisal and the Mexican variety Bobwhite. The later was bred in environments were powdery mildew occurs only in economically insignificant quantities (Lillemo et al. 2006). Therefore, Bobwhite plants did not need powdery mildew resistances in Mexico. This changes however if Bobwhite is planted in the much moister Switzerland.

All experiments showed that the non-GM Bobwhite plants were highly susceptible when grown in Switzerland. The variety Frisal on the other hand, was bred in an environment with strong powdery mildew pressure. After several years of successful cultivation, propagation stopped in 2006 and cultivation in 2008 because of constantly decreasing resistance to fungal pathogens such as powdery mildew. Nevertheless, we found that non-GM Frisal plants proved to be highly resistant to the powdery mildew strains present in our glasshouse and field experiments.

Transgene technology allows introducing fungal resistance genes into conventional wheat varieties such as Bobwhite and Frisal. This should, at least in theory, increase their resistance to powdery mildew and therefore improve seed yields. Introduced resistance genes are usually linked to strong promoters that guarantee much higher expression rates than found in resistance genes of conventional plants. Even if resistance genes originating from cereals are used, we would expect higher resistance than could be achieved with conventional breeding.

We found that lines with *Pm3* transgenes were consistently more resistant to powdery mildew than non-GM control lines (Chapters 1–4). Environmental factors (glasshouse/field, fertilization, competition and climate) influenced the occurrence of powdery mildew and altered the potential benefits of resistance genes. Nevertheless, we could confirm improved resistance in all environments. The five tested GM-Bobwhite lines, *Pm3b*#1, *Pm3b*#2, *Pm3b*#3, *Pm3b*#4 and *Pm3a*#1, differed in their strength of powdery mildew resistance. Line *Pm3b#2* was consistently more resistant than all other *Pm3b* lines (Chapters 1, 2 and 4). The resistance of line *Pm3b*#4 did sometimes but not always reach the levels of *Pm3b*#2 (Chapter 1 and 4) whereas *Pm3b*#1, *Pm3b*#3 and *Pm3a*#1 showed moderate resistance. Quantitative expression data suggests that these differences in resistance can be at least partially explained by different transgene expression levels (Brunner et al. 2011 and personal communications). *Pm3b*#2 was 55 to 617 times higher expressed than in the wheat variety Chul which expresses this gene naturally. All other *Pm3* lines had much lower expression rates. If we rank the GM lines according to their transgene expression level we get the following sequence: *Pm3b*#2 >> *Pm3b*#4 > *Pm3b*#1 > *Pm3b*#3 > *Pm3a*#1. Obviously, there seems to be a correlation between the strength of fungal resistance and the transgene expression rate.

In contrast to the Bobwhite plants, the powdery mildew resistance of GM and non-GM Frisal lines (A9 and A13) did not differ (Chapters 2 and 4). The resistance of the non-GM Frisal plants was quite high. It seems that the overexpression of *Chitinase* or *Chitinase* and *Glucanase* transgenes in the lines A9 and A13 could not further improve the resistance as expected. In Chapter 4 three modern Swiss bread wheat varieties Toronit, Casana and Fiorina were grown alongside our Frisal and Bobwhite lines allowing direct comparisons. We found that GM Frisal lines were equally resistant than the tested Swiss varieties whereas all GM Bobwhite lines were more susceptible.

Summing up, one could argue that the effect of the *Pm3b* transgenes could be demonstrated using highly susceptible Bobwhite plants whereas no improvement could be seen in already resistant Frisal plants with additional *chitinase* and *glucanase* transgenes. It would be of great interest to evaluate the role of the genetic background on the functionality of these transgenes. To do this, the transgenes used in these experiments would need to be cloned into other wheat varieties. Depending on the question asked *Pm3* transgenes could for example be inserted into Frisal plants or *Chitinase* and *Glucanase* transgenes into Bobwhite. Susceptible Bobwhite plants may be useful to study the effectiveness of transgenes, from an applied point of view however, transgenes should be inserted into varieties that are already adapted to a specific agro-environment such as Frisal.

Resistance at high costs

Plant defense in wild plants and crops can lead to reduced performance (Bergelson and Purrington 1996). These so called "fitness costs of resistance" are best measured in absence of a pathogen using GM plants that vary only in individual resistance genes (Purrington 2000; Burdon and Thrall 2003). With fitness costs, ecologists mean the reduction in relative reproductive success (Darwin 1859). In wheat, seed sizes are relatively constant and germination rates close to 100%. Hence, seed yield can be measured to estimate the fitness of individual wheat plants. Seed yield is also one of the most important traits in agronomic assessments. Climatic conditions, nutrient availability and pathogen pressure can influence yields tremendously. Our experiments were performed in different years with low pathogen pressure in 2008 and wetter conditions with more pathogens in 2009. It is therefore not surprising, that seed yields differed between the two years (Chapters 1–4). However, if we look at the differences between GM and non-GM plants, we find relatively consistent patterns. Generally, high pathogen pressure allowed some of our tested GM plants to benefit from their enhanced resistance whereas strong costs of resistance were detected in absence of the pathogens. The yields of the GM Bobwhite lines *Pm3b#2* and *Pm3b#4* were always lower than those of their non-GM sister lines. Even in presence of the pathogen, yields were reduced by half. The performance of *Pm3b#1* changed dramatically with the occurrence of powdery mildew. In absence of this pathogen, the yield of *Pm3b*#1 dropped by 39%. In 2009 however, where powdery mildew spread strongly, this line could increase its yield relative to its non-GM sister line by up to 16%. *Pm3b*#3 and *Pm3a*#1 were not grown in absence of the pathogen. They performed similarly or slightly better than their non-Gm sister lines in presence of the pathogens. The results for *Pm3b*#3 need to be interpreted with care because a parallel study (Brunner et al. 2011) showed that this line suffers from transgene silencing. It is likely that this caused the very high powdery mildew susceptibility of 30% of the *Pm3b*#3 plants grown in our field trials in 2008. It is therefore possible that the overall positive yield effect found in this line was due to individual plants with silenced transgenes that benefited from growing in the vicinity of resistant plants. If we try to rank all GM Bobwhite lines according to their yield we get the following sequence: *Pm3b*#2 > *Pm3b*#4 >> *Pm3b*#1 > *Pm3b*#3 > *Pm3a*#1. As for the powdery mildew resistance, this sequence correlates well the expression level of the transgenes. However, whereas GM lines with high transgene expression could profit from high powdery mildew resistance, it seemed to increase costs of resistance and thus to lower their yield potential.

Although the GM Frisal lines A9 and A13 showed no increased fungal resistance, their yields were lower compared to the conventional variety Frisal (Chapter 2). This was true in presence and absence of powdery mildew. A13 had always lower yields than A9 but differences were only significant in one of two experiments. Nevertheless, one could argue that line A13 which harbours two transgenes could suffer from higher costs of resistance than A9.

We found that the overexpression of *Pm3* and *chitinase* and *glucanase* transgenes can be costly for a plant. For some plant lines, these costs were so large that they could not profit from their enhanced resistance even if pathogens were abundant. Such plants are of no practical use in agriculture. This might be different for stable lines such as *Pm3a*#1 and *Pm3b*#1 whose costs of resistance seem to be smaller than their benefit in environments with high pathogen pressure. We conclude that it is paramount to adjust expression levels in GM plants to levels allowing sufficient pathogen resistance at acceptable costs. Furthermore, advanced promoters that allow reducing the transgene expression to relevant plant tissues or which can be induced in the case of need should be promoted (Brunner et al. 2011). The final point applies not only to GM but also to conventional varieties with costly pathogen resistances. Whereas highly resistant crops might be attractive for farmers due to reduced fungicide input and higher yielding reliability they might lead to lower yields on a global scale. In view of the world food problem it might therefore be safer to cultivate high yielding susceptible alongside with highly resistant cultivars.

Unintended effects and transgene overexpression

In the sections above we described how additional resistance genes can influence powdery mildew resistance and seed yield. We measured, however, a great number of other traits that we did not expect to be linked directly to pathogen resistance. These included fungicide sensitivity, chlorophyll content, stomatal conductance, plant height, seedlings mortality, competitive ability, flowering time, outcrossing rate, flower morphology and infection by different fungi such as ergot. All seven tested GM lines showed several of these unintended effects (Chapters 1–5).

Pm3b lines generally performed poorly if sprayed with fungicides. Besides costs of resistance that were described above, these plants reacted sensitively to fungicide spraying. Their leaves turned yellow which lead to reduced performance. It is possible that the chemical compounds of the transgene interacted somehow with the plants metabolism or the transgene itself. Our field experiments revealed that even untreated plant lines had problems with their leaf physiology. The lines *Pm3b#2* and *Pm3b#3* showed leaf chlorosis which started about two weeks after emergence (Chapter 1; Brunner et al. 2011). This lead to lower chlorophyll contents and reduced stomatal conductance — indicating that the photosynthesis was affected. Again, high transgene expression rates might be responsible for this unintended effect. However, we don't know how fungal resistance genes can influence photosynthesis.

Plant height was found to be lower in the lines *Pm3b#2* and A9. We have currently no explanation for this result. The same is true for seedlings mortality which was higher for all *Pm3b* lines but especially *Pm3b#2* if grown in presence of wireworm larvae. There are at least two possibilities. Either the additional transgenes made these plants more attractive to herbivores or, more likely, reduced their ability to survive herbivore attacks. Reduced herbivore resistance might be an ecological cost caused by the fungal resistance genes. The ability of a plant to compete with its neighbours is of great importance for its ecology. Our phytometer experiments showed that all *Pm3b* lines and A13 had lower competitive performance if grown among other wheat varieties than their non-GM control lines. This may indicate reduced performance in highly competitive environments. However, from the risk aspect, we could expect that the probability that these lines spread and persist in agricultural environments is not greater than or even lower than for non-GM varieties.

Differences in phenological development were found for the lines *Pm3b*#2 and *Pm3a*#1. Both lines developed significantly slower and flowered later than their non-GM control lines. There were also unexpected changes in the flowering behaviour of several lines. Plants with a *Pm3b* transgene were more likely to hybridize with other wheat varieties than non-GM control lines (Chapter 5). This can probably be explained by lower fertility of these lines. Wheat is a strict self pollinator and only a few percent of all flowers are usually pollinated by foreign pollen. We observed however, that the flower morphology of the lines *Pm3b*#2 and *Pm3b*#4 was altered visibly (Chapter 1; Brunner et al. 2011). Flowers stayed open for several days, instead of minutes (De Vries 1971), presumably to allow external pollination. This may indicate that either the pollen quantity or quality was negatively affected (Enjalbert et al. 1998) by the presence of a *Pm3b* transgene. Seeds of the lines *Pm3b*#2, *Pm3b*#3 and *Pm3b*#4 were also significantly more often infected with ergot fungi. It is possible that there is a connection between this infection and the longer flowering period of these plants. Experiments have shown that male-sterile wheat varieties that need to open their flowers to receive foreign pollen are also more prone to ergot infections (Waines and Hegde 2003).

Our results suggest that single transgenes conferring pathogen resistance to plants can lead to unintended effects. To explain why such effects occur is very difficult an unpredictable. However, one of the reasons could be the enhanced gene expression levels in the transgenes which was sometimes several hundred times higher than found in conventional varieties. Indeed, the majority of the unintended effects were found in the lines *Pm3b*#2 and *Pm3b*#4 which have the highest transgene expression. We suggest that altered regulation of single transgenes in plants can have much larger effects than typically observed in the wild. There is definitely need for more research on this topic.

Now that we found these unintended effects we would like to discuss whether they pose a risk to the environment or human health. Most of the effects lead to lower fitness of the study plants. Lower fitness is mainly risky for the plant itself but not for its surrounding. Differences in flowering behavior, especially increased cross-pollination rates could however influence the coexistence of GM and non-GM agriculture and need to be considered. This is of course different for increased ergot infection. Ergot is one of the most toxic fungal pathogens and can cause severe poisoning and paralysis if consumed (Caporael 1976; Fitzhugh et al. 1944). Do this increased ergot infection levels pose a risk to human or animal health? Risk is commonly defined as the product of hazard and exposure (Poppy 2004). We found that in 2008 ergot levels for the lines *Pm3b*#2, *Pm3b*#3 and *Pm3b*#4, were 19, 4, and 20 times higher than the official threshold level (Schweizer Eidgenossenschaft 2010). Such high level would therefore pose a serious health hazard if consumed. However, no feeding experiments were carried out and all seed material was destroyed after the end of the experiments. The exposure was therefore zero. It has to be stated, however, that the finding of increased ergot infection rates was not anticipated and purely coincidental. Nevertheless, it would be highly unlikely that the experimental wheat lines we used would be used in breeding programs since they showed other unintended effects such as reduced yield and necrotic leaves that are easily observed. Ergot infections were not observed in 2009 but again in 2010 (unpublished data). This indicates that environmental and climatic factors can influence ergot infection levels (Fitzhugh et al. 1944). Several years of field trials would therefore be necessary to exclude the possibility of increased ergot infections. However, even in the worst case that increased ergot infection would have gone unnoticed by plant breeders, this would not threaten human health. Ergot is a well known disease and grain lots are checked automatically for ergot kernels. Contaminated grain can be cleaned using sieving techniques which inevitably reduces the overall seed yield. Even in these worst case scenarios, consumers would therefore not be exposed to ergot poisoning. Nevertheless, we can conclude, that the spikes produced by the GM lines *Pm3b*#2, *Pm3b*#3 and *Pm3b*#4 would not be safe for consumption. I did not expect to find such an array of surprisingly large unintended effects prior to our field experiment. Hence, my findings demonstrate how important open and unbiased ecological assessments of GM plants are.

Transgene x environment interactions in GM plants

The study of genotype x environment (G x E) interactions is one of the main topics of ecology and evolutionary biology (Via and Lande 1985). Genotypic expression of traits can differ across environments resulting in large phenotypic variability. G x E interactions are known to be large in wild plants (Schlichting 1986; Sultan 1987; Schmid 1992; Sultan 2001; Yahiaoui et al. 2004) and of great interest for crop scientists (Paterson et al. 2003). The extent to which G x E interactions affect a trait is an important determinant of the amount of testing required over years and locations to satisfactorily quantify the performance of a crop genotype (Paterson et al. 2003). Transgenic plants that differ only in one or a few genes from non-transgenic control lines are particularly valuable for G x E studies. GM plants allow studying how interactions between a particular transgene and environmental factors affect the genotype of a plant.

Our experiments revealed that environmental factors can influence the behaviour of GM and non-GM plants differentially. Moist and relatively warm climatic conditions in 2009 almost doubled yields of our test plants compared to 2008, where it rained less. However, the pathogen powdery mildew could also benefit from moist conditions and caused more severe infections in 2009. Hence, the performance of GM plants resistant to powdery mildew increased relative to highly susceptible non-GM plants (Chapters 1 and 4 vs. Chapters 2 and 3). Similar effects were observed in plots with additional fertilization. Better nutrient availability increased the performance of all plants. However, powdery mildew could also profit from these conditions and infected plants in fertilized plots more than in non-fertilized control treatments. Plant scientists and crop breeders have made similar observations (Chen et al. 2007; Last 1953; Bainbridge 1974; Shaner and Finney 1977). Powdery mildew infections became more serious after the green revolution — mainly because of increased use of nitrogen fertilizer, growth regulators, increased genetic uniformity of crops and increased irrigation (Bennett 1984). Hence, resistant GM plants did profit more from fertilization than non-resistant control plants. Generally, beneficial climatic conditions and nitrogen fertilization increased phenotypic differences between GM and non-GM control lines.

However, the largest transgene x environment interactions were detected when we compared the performance of GM and non-GM plants grown in the glasshouse and in the field. GM plants performed much better in the glasshouse than in the field. More stressful conditions could have increased costs of resistance in GM lines (Chapter 1). Besides the general performance, several unintended effects were only detected in the field. This may have implications for environmental risk assessments.

Substantial equivalence vs. precautionary principle

It has been argued that if GM plants grown in the glasshouse differ from their non-GM parent lines only by expressing the transgene product, the ecological risks of this product can be better tested there than in the field (Raybould 2006; Raybould 2010). This view is based on the principle of "substantial equivalence", a regulatory framework introduced by the OECD in 1993 and later affirmed by the FAO and WHO (OECD 1993; FAO/WHO 1996) aiming to control the introduction of genetically modified crops. Following this principle, a GM plant can be brought to market if one can demonstrate that key toxic or allergenic compounds, key nutrients and possible inherent plant toxins and antinutrients lie within the natural variation found in its natural antecedent (EC 1997). These data are usually gained from plants grown under standardized condition in the laboratory and information about potential transgene × environment interactions are not required. However, from evolutionary and ecological studies on wild plants it is well known that genotype × environment interactions are potentially very large (Schlichting 1986, Sultan 1987, Schmid 1992, Sultan 2001), suggesting that similar interactions might occur in GM plants exposed to different environments, including glasshouse vs. field environments.

Our results show clearly, that introduction of a single transgene influenced the plants performance, physiology and ecology in various ways. Hence, GM plants are not just conventional varieties with some added traits. Whereas the transgene expression and therefore the intended pathogen resistance was only little influenced by environmental conditions, this was not true for unintended effects. When we moved our study plants from the glasshouse to the field their performance and ecology changed. Unintended effects such as reduced yield and plant height, chlorotic leaves, changes in flowering behaviour and increased susceptibility to herbivore and fungal pathogens were not visible in the glasshouse and could only be revealed when plants were grown in realistic field environments. Hence, whereas glasshouse experiments can be useful to study transgene expression and toxicity on non-target organisms, they do not allow to evaluate complex transgene x environment interactions. We argue that information about transgene × environment interactions must be required as a basis for ecological risk assessment. GM plants should be tested in multiple environments that simulate conditions found in agricultural environments. Assessments that are based on the principle of substantial equivalence, which requires only the study of the chemical composition of plants grown in the laboratory, are thus prone to overlook potentially important transgene × environment interactions. It is therefore not surprising that this principle has caused a lengthy debate within the scientific community (Millstone et al. 1999; Novak and Haslberger 2000; Gasson and Burke 2001) and has thus been partly extended or even been replaced by the equally controversial precautionary approach (EU 2000; Barnett et al. 2001) in some countries. Our results show that the value of risk assessments based on the principle of substantial equivalence is limited and we propose to develop and use more sophisticated testing systems (Kuiper et al. 2001; Andow and Hilbeck 2004) that take into account potential transgene × environment interactions.

Value of diversity in agriculture

Biodiversity experiments with wild plants showed, that higher plant diversity increases the total biomass and yield of a system (Tilman et al. 1996; Hector et al. 1999; Roscher et al. 2005). Diverse systems in which different plant species or varieties share different pathogen resistances can slow down spreading of pathogens (Wolfe 1985; Maron et al. 2011). This reduces the pathogen load and makes it more difficult for pathogens to overcome the plants resistances. Mixtures of crop varieties have been cultivated in the past with some success (Smithson and Lenne 1996; Zhu et al. 2000). Although yields of mixed varieties might be higher, they are often less homogeneous than in monocultures which may be seen as a disadvantage. Our aim was to test if GM plants that differ only in few resistance genes but are otherwise phenotypically similar can help to overcome these difficulties. We carried out two experiments where we mixed different GM and non-GM lines with each other (Chapters 2 and 4).

We found no differences for mixtures with GM lines of the variety Frisal, presumably because GM and non-GM control lines were similarly resistant to powdery mildew. It is however not surprising, that mixtures of already resistant lines did not improve resistance further. Mixtures of GM lines with *Pm3* transgenes yielded more interesting results. Resistance to powdery mildew generally increased with increasing GM concentration. Mixtures with equal concentrations of resistant GM and susceptible non-GM sister lines improved the overall resistance which is in line with earlier findings. Furthermore, we found positive diversity effects when GM lines with different resistance alleles were mixed. Mixtures of *Pm3a*#1 and *Pm3b*#1 lines had reduced powdery mildew infection and improved yields compared with monocultures of these lines. In fact, these mixtures provided on of the rare cases of transgressive overyielding in agriculture (Trenbath and Harper 1974; Harper 1977; Vandermeer 1989). In this case, we were lucky to combine two lines with relatively low costs of resistance and complementary resistance genes. However, when we mixed *Pm3b*#1 with *Pm3b*#2 we found no diversity effect. This indicates that not all mixtures will lead to improved resistance or even transgressive overyielding. Plant breeders that developed variety mixtures made similar observations. It took them many years to find optimal mixture compositions (M. Winzeler, personal communications). Nevertheless, mixtures of conventional varieties but also of GM lines have the potential to improve today's agricultural systems.

Unfortunately, the current development leads in the opposite direction. Monocultures of crops become again more common because of international norms asking for homogeneous products (Esquinas-Alcazar 2005). Up to date, few GM varieties with even fewer resistance genes are planted on ever growing areas. This is mainly due to the monopolization of the marked by few large agrochemical companies (Altieri 2000) and strict government regulations (Levidow et al. 2005). The newest trend is stacking several transgenes into individual plants (Halpin 2005). It is possible that the stacking of multiple transgenes within a plant can increase the evolution of resistant pathogens or weeds (Krupke et al. 2009). Furthermore, it is not clear what happens on the population level if several transgenes with resistance costs are combined in a single plant. We would hypothesize that mixtures of resistant plant lines would yield more than monocultures with multiple resistance genes in each plant. However, almost no research has been done in this direction.

I conclude that the almost forgotten agricultural technique of mixing plant varieties or lines should be reconsidered. Biodiversity strategies could help to make our agricultural systems more sustainable and increase productivity at the same time. This may apply to both GM and non-GM cropping systems.

Gene flow and coexistence of GM and non-GM wheat

Assessments of risks and benefits of GM crops usually focus on effects on non-target organisms, health issues and whether or not the performance increases as intended. Such information can help to decide if a particular GM line is safe to be released to the environment or later to be used in commercial cultivation. Crops that do not meet these standards will be rejected. There is, however, a problem that cannot be solved so easily. All sexually reproducing plants can spread their genes through gene flow. Genes can be transferred via pollen among plants of the same species or sometimes even between closely related species. Furthermore, plants can disperse and persist in the form of seeds or seedlings. There are no indications that GM crops differ from non-GM cultivars in the ability to spread their genes (Ellstrand 2003). Some attempts have been made to reduce gene flow of GM crops through "terminator" technology or chloroplastic transformation (Niiler 1999). These efforts were, however, not well received because of various reasons. Seeds from such plants could for example not be used again for sowing the next-year crop (Niiler 1999). Therefore, we can assume that all currently planted GM crops are as fertile as non-GM varieties.

Some proponents of GM technology argue that gene flow among crops and from crops to wild species has happened ever since the beginning of agriculture and caused no harm to human health or the environment (Poppy 2004). Nevertheless, GM policies in most countries demand that gene flow from GM crops need to be managed to keep levels of transgenic contamination below certain thresholds in conventional cops (Devos et al. 2005; Beckie and Hall 2008). Unintended gene flow from GM crops not only enables the introduction into ecosystems of genes that confer novel fitness-related traits, but also allows novel genes to be introduced into many diverse types of crops, each with its own specific potential to outcross (Snow 2002). Furthermore, serious economic losses can be expected if originally conventional fields are contaminated with GM plants (Messean et al. 2007). Transgene escape is likely to happen and can occur though pollen- or seed-mediated gene flow (Ellstrand 2003; Dietiker et al. 2011). Gene flow differs strongly between crop species. Whereas pollen from maize was shown to fertilize plants hundreds of meters away from the pollen source at high rates (Devos et al. 2005; Mallory-Smith and Zapiola 2008) this is different for wheat. Wheat is a prominently self-pollinating species which means that flowers are usually pollinated by pollen of the same plant. Nevertheless, some plants are cross-pollinated (Enjalbert et al. 1998) which leads gene flow (Gatford et al. 2006; Matus-Cádiz et al. 2007). Knowledge about gene flow of GM wheat is therefore important to allow coexistence of GM and non-GM wheat crops in the future.

Our data show that already short isolation distances of about 2 meters are sufficient to separate GM from non-GM wheat varieties. We could also confirm that cross-pollination in wheat is relatively low (3.4%) even among neighbouring plants. However, in contrast to our intuition, we found that the insertion of a transgene can influence both the plants' fertility and cross-pollination rates. GM plants with *Pm3b* transgenes were less fertile and 6 times more likely to receive foreign pollen than non-GM lines.

Pollen flow is not the only way transgenes can spread in an agro-environment. Studies have shown that it might be more likely that seed admixture occurs already before sowing because seed production chains of GM and non-GM seeds cannot be separated completely (Dietiker et al. 2011). We therefore used our cross-pollination data to calculate how much GM plants would spread if a conventional wheat field would be contaminated with 0.9% GM wheat seeds. We found that by the time of harvest, the proportion of seeds with at least one copy of the transgene would increase by 0.031%. Hence, we would not expect transgenes to spread fast within contaminated wheat fields. This would, however, be different in outcrossing crops such as barley or maize. We conclude that pollen- and seed-mediated gene flow of GM wheat can be managed using short isolation distances and adequate threshold limits. However, our results also show that transgenes can alter the flowering biology and influence pollen- and seed-mediated gene flow. This suggests that data from conventional crop varieties might not be reliable enough to calculate isolation distances and gene-flow management schemes. It could be safer to assess the gene flow potential of GM crops using a case-by-case approach.

Innovative methods for GM crop assessments and agricultural research

The performance of wheat is usually assessed by agronomist using relatively large field trials. However, we were not only interested in basic agricultural variables such as yield, seed set and pathogen susceptibility but also in how crop-plant traits change depending on the environment. Hence, complex experimental designs and special techniques, commonly used by ecologists to study wild plants, were necessary to assess such interactions.

The performance and ecology of plants can be studied either at the level of the individual plant or the population. In agronomy, population measures are most important because the performance of plants needs to be put in relation to the area it takes to grow them. Nevertheless, our results show (Chapter 3) that data gained from individual plants can be used to predict the behaviour of the entire population. Furthermore, fitness-related traits, pathogen incidence, phenological development and interactions among plants can be better studied at the level of individual plants. Hence, we propose to use both approaches. We believe that combining data on individual plants and entire populations can lead to better understanding of agricultural systems and improve assessments of GM plants.

One of these methods is called "phytometer technique" (Chapter 4). Phytometers are individual plants, transplanted into a range of biotic and abiotic environments. This technique was developed to measure the quality of environments but can also be used to assess the response of different species grown in different environments (Mwangi et al. 2007). To our knowledge, the phytometer technique so far has never been used to study GM plants. Using this technique allowed us to assess the performance and competitive ability of 15 different GM and non-GM lines grown simultaneously in 15 biotic and two abiotic environments on less than 130 m^2 of field plots (see discussion above). Conclusive data could be generated on very limited space helping to keep costs of field trials low. We are therefore confident that the phytometer technique has a high potential to be used in future assessments of GM plants but also in applied agricultural research in general.

Phytometers can also be used to measure cross-pollination rates within a field (Chapter 5). We planted individual GM and non-GM plants of the variety Bobwhite into plots of different varieties like Frisal and Casana. The later were chosen because we observed that crosses between these varieties and Bobwhite lines suffer from necrotic leaf tips. Using this visual sign as marker we could easily and reliably identify cross-pollinated offspring saving us thousands of PCR reactions. We propose to use this approach to gain basic gene flow data on partially self-pollinating GM and non-GM crops.

Finally, we used an innovative approach to assess pollen-mediated gene flow over short distances. Flour produced from batches of 100 seeds was tested for presence or absence of transgenes. Maximum likelihood estimation for binomial data (Fischer 1922) was then used to calculate most likely cross-pollination rates. This approach allowed us to measure even tiny transgenic contaminations of 0.02% reliably and quickly. We propose therefore to use this population-based PCR approach to analyse gene flow between GM and non-GM crops at least until more sophisticated quantitative RT-PCR methods are available.

Relevance for plant biotechnology

Four years of glasshouse experiments and field trials with experimental GM wheat varieties have yielded mixed results. Pathogen resistance of the variety Bobwhite could be improved trough the introduction and overexpression of *Pm3* transgenes. However, additional *chitinase* and *glucanase* transgenes did not increase resistance to fungal pathogens in the variety Frisal. GM lines performed worse due to high costs of resistance in absence and partly even in presence of the pathogen. These costs were highest for lines with high transgene expression or in lines with more than one transgene. Furthermore, a variety unintended effects that affected the plants' morphology, phenology and ecology were found in lines with particularly strong transgene expression.

Our results suggest that even after years of research, the production of GM lines that express the intended traits and nothing else seems to be particularly difficult and involves random processes. We found that GM lines with similar transgenes and identical genetic background differed greatly from each other possibly due to positional effects that altered the transgene expression. In some cases, this expression seemed to be too high for the plant to cope with. This may indicate that differing transgene expression rates influence the performance and ecology of a plant more than we could expect from the presence or absence of an additional genes. Since every transformed plant coped differently with inserted transgenes we propose that GM lines need to be assessed on a case-by-case basis (Andow and Zwahlen 2006). This applies also to environmental risk assessments. We do not believe that the safety of a particular transgene can be guaranteed independently of the organism it is inserted in, as proclaimed by some scientists (De Schrijver et al. 2007).

Furthermore, transgene x environment interactions were shown to be very large. Plants performed very differently when cultivated in the glasshouse than in the field. Hence, we suggest that, whenever possible, assessments of the performance and risks of GM plants should be carried out directly in the field. Laboratory and glasshouse experiments alone might not allow identifying and excluding plants with unintended and potentially harmful effects before their release to the environment. Hence, the safety of such field trials needs to be guaranteed to minimize the risk of transgene escapes. This problem seems to be manageable at least for GM wheat. We found that cross-pollination rates are generally low in wheat and gene flow happens only over very short distances.

We conclude that the experimental GM lines with which we worked differed from their non-GM sister or control lines in many traits. Some of them were of ecological or even health-related relevance. Well-designed field trials are therefore essential to exclude lines with unintended traits and identify potentially useful ones. Many of the problems encountered with these experimental wheat lines can also be observed in new varieties produced by conventional breeding. Only plant breeders can breed novel genetic information into varieties that are optimally adapted to the harsh conditions of their agro-ecosystem. I believe that whereas biotechnology might provide plants with new traits it cannot and should not replace traditional plant breeding.

Will biotechnology solve the world food problem?

Proponents of GM technology proclaimed repeatedly that GM crops will allow increasing food production to meet the demand of an ever growing global population (Conway and Toenniessen 1999; Khush 2001; James 2009). Some go so far to allege GM critics to be responsibly for today's famines (Borlaug 2000). However, up to now, there is little evidence that increased yields are caused by GM crops and not just improved agricultural practice (Qaim and Zilberman 2003; IAASTD 2009; Sheridan 2009). Our results show, that crops with additional transgenes are likely to suffer from costs of resistance. Such trade-offs have been well studied by ecologists and are to be expected (Bergelson and Purrington 1996). Only the future will show if, in the absence of pathogens, absolute yield *gains* of GM crops with resistance genes will ever be possible. However, GM crops could potentially reduce yield *loss* due to better pathogen and abiotic stress resistance under reduced input of agrochemicals. This could make agriculture more sustainable and lower the variation in crop performance which increases food safety. However, today's monopolisation of the applied biotechnology research by few agrochemical companies could threaten this goal (Godfray et al. 2010; Serageldin 1999). Genetic diversity of important crops is declining fast (Esquinas-Alcazar 2005). It is conceivable that the widespread cultivation of relatively similar GM crops supports this trend. However, reduced crop diversity can lower the effectiveness of basic ecosystems services such as pest and disease management, pollination and soil processes (Hajjar et al. 2008), as well as the natural resources of new genes for biotechnological applications.

Basic ecological research could help to understand and eventually manage invading pathogens (Tilman et al. 1996). Diversification strategies could increase the complexity of monotonous agricultural fields, thus reducing their susceptibility to pathogen epidemics (Pimentel 1991). Our results show (Chapter 3) that mixtures of experimental GM lines can reduce pathogen incidence and increase yields transgressively. Even better results might be achieved if GM varieties are combined that have already gone trough a breeding process. We propose that GM and non-GM farming systems could benefit from the use of diversification strategies. There are also ideas how to reduce yield losses due to weeds. Ecologists have shown that increasing cropping density and improved sowing pattern can reduce the growth of weeds without input of chemicals or additional work (Weiner et al. 2010). They also propose novel breeding techniques that aim to improve the performance of entire crop populations. These examples show that basic ecological principles could be of great use to applied agronomy. If promoted, so called evolutionary agro-ecology (Weiner et al. 2010) could also help to mitigate the world food crisis.

Back to the question asked at the beginning of this section: will GM crops solve the world food crisis? I believe that GM crops could potentially help to increase food production if used more carefully than today. However, biotechnology can at best contribute partially to the solution of the problem. Better use of already available ecological knowledge does also hold the potential to revolutionize agriculture. To overcome the massive problems the world is facing today, researchers of biotechnology, agronomy and ecology need join their forces and work towards a sustainable future.

References

Altieri, M. A. (2000) The ecological impacts of transgenic crops on agroecosystem health. *Ecosystem Health,* **6,** 13-23.

Andow, D. A. & Hilbeck, A. (2004) Science-based risk assessment for nontarget effects of transgenic crops. *Bioscience,* **54,** 637-649.

Andow, D. A. & Zwahlen, C. (2006) Assessing environmental risks of transgenic plants. *Ecology Letters,* **9,** 196-214.

Bainbridge, A. (1974) Effect of nitrogen nutrition of host on barley powdery mildew. *Plant Pathology,* **23,** 160-161.

Barnett, S., Bware-Rogers, J., Brunk, C., Caulfield, T., Ellis, B., Fortin, M. & al., e. (2001) *Elements of precaution: Recommendations for regulation of food biotechnology in Canada.* The royal society of Canada, Ottawa, Canada.

Beckie, H. J. & Hall, L. M. (2008) Simple to complex: modelling crop pollen-mediated gene flow. *Plant Science,* **175,** 615-628.

Bennett, F. G. A. (1984) Resistance to powdery mildew in wheat: a review of its use in agriculture and breeding programmes. *Plant Pathology,* **33,** 279-300.

Bergelson, J. & Purrington, C. B. (1996) Surveying patterns in the cost of resistance in plants. *American Naturalist,* **148,** 536-558.

Borlaug, N. E. (2000) Ending world hunger. The promise of biotechnology and the threat of antiscience zealotry. *Plant Physiol.,* **124,** 487-490.

Brunner, S., Hurni, S., Herren, G., Kalinina, O., von Burg, S., Zeller, S., Schmid, B., Winzeler, M. & Keller, B. (2011) Transgenic *Pm3b* wheat lines show resistance to powdery mildew in the field. *Plant Biotechnology Journal,* **in press**.

Burdon, J. J. & Thrall, P. H. (2003) The fitness costs to plants of resistance to pathogens. *Genome Biology,* **4,** 1-3.

Caporael, L. R. (1976) Ergotism: the satan loosed in Salem? *Science,* **192,** 21-26.

Chen, Y. X., Zhang, F. D., Tang, L., Zheng, Y., Li, Y. J., Christie, P. & Li, L. (2007) Wheat powdery mildew and foliar N concentrations as influenced by N fertilization and belowground interactions with intercropped faba bean. *Plant and Soil,* **291,** 1-13.

Conway, G. & Toenniessen, G. (1999) Feeding the world in the twenty-first century. *Nature,* **402,** 55-58.

Darwin, C. (1859) *On the origin of species by means of natura selection, or the preservation of favoured races in the struggle for life.* Atheneum, New York.

De Schrijver, A., Devos, Y., Van den Bulcke, M., Cadot, P., De Loose, M., Reheul, D. & Sneyers, M. (2007) Risk assessment of GM stacked events obtained from crosses between GM events. *Trends in Food Science & Technology,* **18,** 101-109.

De Vries, A. P. (1971) Flowering biology of wheat, particularly in view of hybrid seed production - review. *Euphytica,* **20,** 152-170.

Devos, Y., Reheul, D. & De Schrijver, A. (2005) The co-existence between transgenic and non-transgenic maize in the European Union: a focus on pollen flow and cross-fertilization. *Environmental Biosafety Research,* **4,** 71-87.

Dietiker, D., Oehen, B., Ochsenbein, C., Westgate, M. E. & Stamp, P. (2011) Field simulation of transgenic seed admixture dispersion in maize with a blue kernel colour marker. *Crop Science,* **51,** doi: 10.2135/cropsci2010.06.0311.

EC (1997) Directive (EC) No. 258/97 of the European Parlament and Council (Jan. 27 1997) on novel food and novel food components. *Official Journal of the Europiean Communities,* **43,** 1-6.

Ellstrand, N. C. (2003) Current knowledge of gene flow in plants: implications for transgene flow. *Philosophical Transactions of the Royal Society B-Biological Sciences,* **358,** 1163-1170.

Enjalbert, J., Goldringer, I., David, J. & Brabant, P. (1998) The relevance of outcrossing for the dynamic management of genetic resources in predominantly selfing *Triticum aestivum* L. (bread wheat). *Genetics Selection Evolution,* **30,** 197-211.

Esquinas-Alcazar, J. (2005) Protecting crop genetic diversity for food security: political, ethical and technical challenges. *Nature Reviews Genetics,* **6,** 946-953.

EU (2000) *Communication on the precautionary principle.* Commission of the European communities, Brussels.

FAO/WHO (1996) Joint FAO/WHO expert consultation on biotechnology and food safety. Rome, Italy, 30. September to 4 October 1996. ftp://ftp.fao.org/es/esn/food/biotechnology.pdf.

Fitzhugh, O. G., Nelson, A. A. & Calvery, H. O. (1944) The chronic toxicity of ergot. *Journal of Pharmacology and Experimental Therapeutics,* **82,** 364-376.

Gasson, M. & Burke, D. (2001) Scientific perspectives on regulating the safety of genetically modified foods. *Nature Reviews Genetics,* **2,** 217-222.

Gatford, K., Basri, Z., Edlington, J., Lloyd, J., Qureshi, J., Brettell, R. & Fincher, G. (2006) Gene flow from transgenic wheat and barley under field conditions. *Euphytica,* **151,** 383-391.

Godfray, H. C. J., Beddington, J. R., Crute, I. R., Haddad, L., Lawrence, D., Muir, J. F., Pretty, J., Robinson, S., Thomas, S. M. & Toulmin, C. (2010) Food security: the challenge of feeding 9 billion people. *Science,* **327,** 812-818.

Hajjar, R., Jarvis, D. I. & Gemmill-Herren, B. (2008) The utility of crop genetic diversity in maintaining ecosystem services. *Agriculture Ecosystems & Environment,* **123,** 261-270.

Halpin, C. (2005) Gene stacking in transgenic plants - the challenge for 21st century plant biotechnology. *Plant Biotechnology Journal,* **3,** 141-155.

Harper, J. L. (1977) *Population biology of plants.* Academic Press, London.

Hector, A., Schmid, B., Beierkuhnlein, C., Caldeira, M. C., Diemer, M., Dimitrakopoulos, P. G., Finn, J. A., Freitas, H., Giller, P. S., Good, J., Harris, R., Hogberg, P., Huss-Danell, K., Joshi, J., Jumpponen, A., Korner, C., Leadley, P. W., Loreau, M., Minns, A., Mulder, C. P. H., O'Donovan, G., Otway, S. J., Pereira, J. S., Prinz, A., Read, D. J., Scherer-Lorenzen, M., Schulze, E. D., Siamantziouras, A. S. D., Spehn, E. M., Terry, A. C., Troumbis, A. Y., Woodward, F. I., Yachi, S. & Lawton, J. H. (1999) Plant diversity and productivity experiments in European grasslands. *Science,* **286,** 1123-1127.

IAASTD (2009) *International assessment of agricultural knowledge, science and technology for development - synthesis report.* Island Press, Washington DC.

James, C. (2009) *Global status of commercialized biotech/GM crops: 2009.* ISAAA, Ithaca, NY.

Khush, G. S. (2001) Green revolution: the way forward. *Nature Reviews Genetics,* **2,** 815-822.

Krupke, C., Marquardt, P., Johnson, W., Weller, S. & Conley, S. P. (2009) Volunteer corn presents new challenges for insect resistance management. *Agronomy Journal,* **101,** 797-799.

Kuiper, H. A., Kleter, G. A., Noteborn, H. P. J. M. & Kok, E. J. (2001) Assessment of the food safety issues related to genetically modified foods. *Plant Journal,* **27,** 503-528.

Last, F. T. (1953) Some effects of temperature and nitrogen supply on wheat powdery mildew. *Annals of Applied Biology,* **40,** 312-322.

Levidow, L., Carr, S. & Wield, D. (2005) European Union regulation of agri-biotechnology: precautionary links between science, expertise and policy. *Science and Public Policy,* **32,** 261-276.

Lillemo, M., Skinnes, H., Singh, R. P. & van Ginkel, M. (2006) Genetic analysis of partial resistance to powdery mildew in bread wheat line Saar. *Plant Disease,* **90,** 225-228.

Mallory-Smith, C. & Zapiola, M. (2008) Gene flow from glyphosate-resistant crops. *Pest Management Science,* **64,** 428-440.

Maron, J. L., Marler, M., Klironomos, J. N. & Cleveland, C. C. (2011) Soil fungal pathogens and the relationship between plant diversity and productivity. *Ecology Letters,* **14,** 36-41.

Matus-Cádiz, M. A., Hucl, P. & Dupuis, B. (2007) Pollen-mediated gene flow in wheat at the commercial scale. *Crop Science,* **47,** 573-579.

Messean, A., Sausse, C., Gasquez, J. & Darmency, H. (2007) Occurrence of genetically modified oilseed rape seeds in the harvest of subsequent conventional oilseed rape over time. *European Journal of Agronomy,* **27,** 115-122.

Millstone, E., Brunner, E. & Mayer, S. (1999) Beyond 'substantial equivalence'. *Nature,* **401,** 525-526.

Mwangi, P. N., Schmitz, M., Scherber, C., Roscher, C., Schumacher, J., Scherer-Lorenzen, M., Weisser, W. W. & Schmid, B. (2007) Niche pre-emption increases with species richness in experimental plant communities. *Journal of Ecology,* **95,** 65-78.

Niiler, E. (1999) Terminator technology temporarily terminated. *Nature Biotechnology,* **17,** 1054-1054.
Novak, W. K. & Haslberger, A. G. (2000) Substantial equivalence of antinutrients and inherent plant toxins in genetically modified novel foods. *Food and Chemical Toxicology,* **38,** 473-483.
OECD (1993) Safety evaluation of foods derived by modern biotechnology: concepts and principles. Organisation for Economic Co-operation and Development (OECD), Paris.
Paterson, A. H., Saranga, Y., Menz, M., Jiang, C. X. & Wright, R. J. (2003) QTL analysis of genotype x environment interactions affecting cotton fiber quality. *Theoretical and Applied Genetics,* **106,** 384-396.
Pimentel, D. (1991) Diversification of biological-control strategies in agriculture. *Crop Protection,* **10,** 243-253.
Poppy, G. M. (2004) Geneflow from GM plants - towards a more quantitative risk assessment. *Trends in Biotechnology,* **22,** 436-438.
Purrington, C. B. (2000) Costs of resistance. *Current Opinion in Plant Biology,* **3,** 305-308.
Qaim, M. & Zilberman, D. (2003) Yield effects of genetically modified crops in developing countries. *Science,* **299,** 900-902.
Raybould, A. (2006) Problem formulation and hypothesis testing for environmental risk assessments of genetically modified crops. *Environ. Biosafety Res.,* **5,** 119-125.
Raybould, A. (2010) Reducing uncertainty in regulatory decision-making for transgenic crops - More ecological research or clearer environmental risk assessment? *Landes Bioscience,* **1,** 1-7.
Roscher, C., Temperton, V. M., Scherer-Lorenzen, M., Schmitz, M., Schumacher, J., Schmid, B., Buchmann, N., Weisser, W. W. & Schulze, E. D. (2005) Overyielding in experimental grassland communities - irrespective of species pool or spatial scale. *Ecology Letters,* **8,** 576-577.
Schlichting, C. D. (1986) The evolution of phenotypic plasticity in plants. *Annual Review of Ecology and Systematics,* **17,** 667-693.
Schmid, B. (1992) Phenotypic variation in plants. *Evolutionary Trends in Plants,* **6,** 45-60.
SchweizerEidgenossenschaft (2010) Verordnung des EDI über Fremd- und Inhaltsstoffe in Lebensmitteln: Liste der zugelassenen Höchstkonzentrationen (Toleranz- und Grenzwerte) für Pflanzenschutzmittel, Vorratsschutzmittel sowie Regulatoren für die Pflanzenentwicklung. www.admin.ch/ch/d/sr/817_021_23/app1.html.
Serageldin, I. (1999) Biotechnology and food security in the 21st century. *Science,* **285,** 387-389.
Shaner, G. & Finney, R. E. (1977) The effect of nitrogen fertilization on the expression of slow-mildewing resistance in Knox wheat. *Purdue University Agricultural Experiment Station Series,* 1051-1056.
Sheridan, C. (2009) Report claims no yield advantage for Bt crops. *Nature Biotechnology,* **27,** 588-589.
Smithson, J. B. & Lenne, J. M. (1996) Varietal mixtures: A viable strategy for sustainable productivity in subsistence agriculture. *Annals of Applied Biology,* **128,** 127-158.
Snow, A. A. (2002) Transgenic crops - why gene flow matters. *Nature Biotechnology,* **20,** 542-542.
Sultan, S. E. (1987) Evolutionary implications of phenotypic plasticity in plants. *Evolutionary Biology,* **21,** 127-178.
Sultan, S. E. (2001) Phenotypic plasticity for fitness components in *Polygonum* species of contrasting ecological breadth. *Ecology,* **82,** 328-343.
Tilman, D., Wedin, D. & Knops, J. (1996) Productivity and sustainability influenced by biodiversity in grassland ecosystems. *Nature,* **379,** 718-720.
Trenbath, B. R. & Harper, J. L. (1974) Neighbor effects in genus *Avena*. 2. Comparison of weed species. *Journal of Applied Ecology,* **11,** 111-125.
Vandermeer, J. (1989) *The ecology of intercropping.* Cambridge University Press, Cambridge.
Via, S. & Lande, R. (1985) Genotype-environment interaction and the evolution of phenotypic plasticity. *Evolution,* **39,** 505-522.

Waines, J. G. & Hegde, S. G. (2003) Intraspecific gene flow in bread wheat as affected by reproductive biology and pollination ecology of wheat flowers. *Crop Science,* **43,** 451-463.
Weiner, J., Andersen, S. B., Wille, W. K. M., Griepentrog, H. W. & Olsen, J. M. (2010) Evolutionary agroecology: the potential for cooperative, high density, weed-suppressing cereals. *Evolutionary Applications,* **3,** 473-479.
Wolfe, M. S. (1985) The current status and prospects of multiline cultivars and variety mixtures for disease resistance. *Annual Review of Phytopathology,* **23,** 251-273.
Yahiaoui, N., Srichumpa, P., Dudler, R. & Keller, B. (2004) Genome analysis at different ploidy levels allows cloning of the powdery mildew resistance gene *Pm3b* from hexaploid wheat. *Plant Journal,* **37,** 528-538.
Zhu, Y., Chen, H., Fan, J., Wang, Y., Li, Y., Chen, J., Fan, J., Yang, S., Hu, L., Leung, H., Mew, T. W., Teng, P. S., Wang, Z. & Mundt, C. C. (2000) Genetic diversity and disease control in rice. *Nature,* **406,** 718-722.
Figure 7: Photograph taken by Andrea Foetzki. Persons on the picture: Simone Nägeli, Silvan Rieben, Olena Kalinina and Simon Zeller (left to rig

SUMMARY

Fig. 7: Field trial at ART Reckenholz with *triticale* border crop in June 2009 (S. Zeller)

The global population and the demand for protein-rich foods and bio-fuels are growing fast. If this trend continues as predicted, world food security will be at risk. Attempts to increase agricultural production will meet severe limitations due to dwindling resources. In the past, improved crop varieties have allowed to increase production on a given area. Proponents of green biotechnology argue that genetically modified (GM) crops could increase yields while reducing input of agrochemicals. However, there are reports showing that GM crops can propagate and persist within and outside of agricultural environments and share their transgenes with other crops or wild species. This unintended gene flow could potentially threaten coexistence of GM and non-GM agricultural systems and might affect ecosystem services. It is therefore essential to understand the ecology of GM plants. Little is known how transgenic plants interact with their environment and generally, how plants cope with additional genes or genes with changed expression levels. Here we performed several glasshouse and field experiments to assess the performance, resistance costs and unintended effects of seven experimental GM wheat lines grown in different nutritional and competitive environments. Furthermore, we studied the influence of transgenes on pollen-mediated gene flow.

We found that five wheat lines of the variety Bobwhite with inserted *Pm3* transgenes were more resistant to the fungal pathogen powdery mildew than non-GM control lines. However, no improvement was observed when either *chitinase* or and *chitinase* and *glucanase* transgenes were inserted into already resistant Frisal plants. Increased fungal resistance did not translate directly into higher yields. In the absence of the pathogen, all tested GM lines performed worse than non-GM control lines. These costs of resistance were highest for lines with the strongest transgene expression or plants with two different transgenes. In environments with high pathogen pressure two lines *Pm3a#1* and *Pm3b#1* performed slightly better than their non-GM control lines whereas the opposite was true for other lines. All tested GM lines showed at least some unintended effects, i.e., hyper-sensitivity to fungicide spraying, chlorotic leaves, reduced plant height, reduced herbivore resistance, lower competitive ability if grown among other wheat varieties and changes in flower morphology. The later might indicate that lines with strong *Pm3b* overexpression suffer from fertility problems, forcing them to open their flowers to allow cross-pollination. This might also explain high hybridisation rates and increased infections by ergot fungi. We found that environmental factors can influence the behaviour of GM and non-GM plants differently. Generally, beneficial climatic conditions and nitrogen fertilization increased these differences. The largest transgene x environment interactions were found between GM and non-GM plants grown in the glasshouse or the field. GM plants performed generally poor and showed unintended effects in the field, whereas they performed well and showed no such effects in the glasshouse. Powdery mildew resistance and yield increased when lines with different *Pm3* alleles were grown in mixtures. It has to be seen if mixtures of relatively similar GM lines perform better than uniform monocultures with several stacked transgenes. Finally, we could confirm that short isolation distances are sufficient to separate GM from non-GM wheat varieties. Whereas cross-pollination rates were generally low, *Pm3b* lines were more likely to hybridize with neighbouring plants than were non-GM controls. Hence, we found that inserted transgenes can alter the reproductive biology and influence pollen- and seed-mediated gene flow.

We conclude that it is still challenging to produce GM crops expressing only intended but no unintended traits. Our results suggest that transgene expressions levels influence the performance and ecology of GM crops more than we would expect due to the presence or absence of additional genes. It is therefore essential to assess risks and benefits of GM lines on a case-by-case basis. Such assessments should include field trials and multiple environmental factors to reveal transgene x environment interactions. In the case of wheat, safe field trials and the coexistence of GM and non-GM wheat are made easier than in other species due to limited pollen-mediated gene flow. We found strong negative trade-offs between transgene expression and plant performance. Although we cannot extrapolate from our results to other GM crops, it is conceivable that many resistance genes are linked to additional costs which reduce the yield potential under certain environmental conditions. On the other hand, diversification strategies were successful, even when experimental GM lines were used. Hence, basic ecological theory can potentially improve the sustainability and productivity of agricultural systems. I believe that major efforts will be necessary to feed the world in the future and biotechnology could be one tool to achieve this if used carefully. However, classical breeding and ecological research should not be neglected because they might also hold the potential to revolutionize today's agriculture.

Fig. 8: Threshed seeds and ergots of a GM-Bobwhite line (*Pm3b*#2) in 2008 (S. Zeller)

Wir leben in einer Zeit, in der die Weltbevölkerung wächst und die Nachfrage nach proteinreichen Nahrungsmitteln und Biotreibstoffen stark zunimmt. Wenn diese Trends wie vorhergesagt anhalten, wird sich die Welternährungslage verschlechtern. Denn schwindende Ressourcen wie z.B. Land, Wasser, fossile Energieträger und Nährstoffe werden sehr wahrscheinlich die Steigerung der landwirtschaftlichen Produktion erschweren. Moderne Pflanzensorten haben in der Vergangenheit erhebliche Produktionssteigerungen ermöglicht. Verfechter der grünen Gentechnik sind überzeugt, dass mithilfe von gentechnisch veränderten (GV) Pflanzen die Erträge erhöht und gleichzeitig der Verbrauch von Pflanzenschutzmitteln reduziert werden können. Anderseits gibt es inzwischen Publikationen, die zeigen, dass sich GV-Pflanzen in landwirtschaftlichen Anbaugebieten vermehren und ihre Transgene an Nutz- und Wildpflanzen weitergeben können. Unkontrollierter Genfluss könnte die Koexistenz von GV und nicht-GV Pflanzen gefährden und das Funktionieren von Agrarökosystemen negativ beeinflussen. Deshalb ist es wichtig, die Ökologie von GV-Pflanzen besser zu verstehen. Bisher weiss man wenig über Interaktionen zwischen GV-Pflanzen und ihrer Umwelt und wie zusätzliche Gene bzw. Veränderungen in der Genexpression das Verhalten von Pflanzen beeinflussen. Daher haben wir mehrere Gewächshaus- und Feldversuche durchgeführt, um das Verhalten von sieben experimentellen GV-Weizen Linien in verschiedenen Nährstoff- und Konkurrenzsituationen zu untersuchen. Zudem untersuchten wir, ob Resistenzkosten, unerwünschte Nebeneffekten oder Veränderungen im Genfluss vermehrt in GV-Pflanzen auftreten.

Unsere Resultate zeigen, dass fünf Weizenlinien der Sorte Bobwhite, die ein zusätzliches Resistenzgen (*Pm3*) erhalten haben, weniger stark von Mehltaupilzen befallen wurden als nicht veränderte Kontrolllinien. Bei Linien der bereits gegen Mehltau resistenten Sorte Frisal, die über zusätzliche Transgene zur Herstellung von *Chitinasen* oder *Chitinasen* und *Glukanasen* verfügen, konnte jedoch keine erhöhte Resistenz festgestellt werden. Verbesserte Pilzresistenz führte nicht automatisch zu grösseren Erträgen. Wenn keine Mehltauinfektion auftrat, waren die Erträge bei allen untersuchten GV-Linien geringer als bei nicht veränderten Kontrolllinien. Diese sogenannten Resistenzkosten waren bei den GV-Pflanzen mit den stärksten Transgenexpression am grössten. Unter sehr starkem Mehltauinfektionsdruck zeigten zwei GV-Linen, *Pm3a*#1 und *Pm3b*#1, etwas bessere Erträge als ihre nicht veränderten Kontrolllinien. Bei allen anderen GV-Linien waren die Erträge auch unter solchen Bedingungen geringer. Unterwünschte Nebeneffekte konnten bei allen GV-Linien festgestellt werden, wenn auch in unterschiedlicher Ausprägung. GV-Pflanzen reagierten unter anderem besonders sensitiv auf Fungizidbehandlung, hatten chlorotische Blätter, reduzierte Wuchshöhe, geringere Resistenz gegen Herbivore, geringere Konkurrenzkraft und die Blütenmorphologie war verändert. Letzteres trat bei *Pm3b* Linien mit besonders starker Transgenexpression auf und könnte auf Fertilitätsprobleme hindeuten. Obwohl Weizen eigentlich ein Selbstbefruchter ist, mussten diese Pflanzen ihre Blüten lange öffnen um von fremden Pollen bestäubt zu werden. Dies könnte auch die hohe Hybridisationsraten und auffällig starke Mutterkorninfektionen erklären.

Umwelteinflüsse können das Verhalten von GV- und ihren nicht veränderten Kontrollpflanzen unterschiedlich beeinflussen. Im Allgemeinen verstärkten günstige klimatische Bedingungen und Stickstoffdüngung diese Unterschiede. Besonderst starke Interaktionen zwischen dem Transgen und der Umwelt traten auf, wenn GV- und nicht veränderte Pflanzen unter Gewächshaus- oder Feldbedingungen angepflanzt wurden. Im Gewächshaus profitierten praktisch alle GV-Pflanzen von ihrer verbesserten Mehltauresistenz, was sich in höheren Erträgen bemerkbar machte. Im Gegensatz dazu wuchsen GV-Pflanzen im Feld im Allgemeinen schlechter als nicht veränderte Kontrollpflanzen und zeigten unerwünschte Nebeneffekte. Das Mischen von GV-Linien mit verschiedenen *Pm3*-Allelen führte zu verbesserter Mehltauresistenz und höheren Erträgen. Es wird sich zeigen, ob Mischungen mit relativ ähnlichen GV-Linien gegenüber Monokulturen mit multiplen Transgenen im Vorteil sind. Schliesslich konnten wir bestätigen, dass kurze Isolationsdistanzen genügen um GV- von nicht GV-Pflanzen zu trennen. Auskreuzungsraten waren im Allgemeinen gering. GV-Linien mit *Pm3b*-Transgenen kreuzten sich jedoch häufiger mit Nachbarpflanzen als nicht veränderte Kontrolllinen. Dies zeigt, dass Transgene die Reproduktionsbiologie und den Genfluss zwischen Pflanzen beeinflussen können.

Wir schliessen aus unseren Ergebnissen, dass die Herstellung von GV-Pflanzen, die nur die erwünschten Effekte zeigen, weiterhin schwierig ist. Die Expressionsstärke von Transgenen scheint das Verhalten und insbesondere die Ökologie von GV-Pflanzen stärker zu beeinflussen, als man durch das Vorhandensein oder Fehlen von Resistenzgenen erwarten könnte. Demzufolge müssen Risiken und Nutzen jeder neuen GV-Linie separat untersucht werden. Solche Abschätzungen sollten auch Feldversuche und verschiedene Umweltbedingungen beinhalten, damit Interaktionen zwischen dem Transgen und der Umwelt aufgedeckt werden können. Sichere Feldversuche können mit Weizen dank tiefer Auskreuzungsrate leichter durchgeführt werden als mit anderen Pflanzenarten. Die Expression der Transgene korrelierte negativ mit dem Ertrag und anderen wichtigen agronomischen Merkmalen. Obwohl wir aufgrund unserer Resultate nicht auf das Verhalten anderer GV-Pflanzen schliessen können, ist es denkbar, dass Resistenzgene zu zusätzlichen Kosten für die Pflanzen führen, die abhängig von den jeweiligen Umweltbedingungen das Ertragspotential schmälern können. Anderseits war unsere Diversifikationsstrategie, d.h. die Mischung von GV-Linien mit verschiedenen *Pm3*-Allelen, erfolgreich, obwohl diese Mischungen nur zufällig und nicht zielgerichtet zusammengestellt worden waren. Dies zeigt, dass die Anwendung von grundlegenden Theorien aus der Ökologie die Produktivität und Nachhaltigkeit von Agrarsystemen verbessern könnte.

Ich bin zur Überzeugung gekommen, dass massive Anstrengungen notwendig sind, damit die bereits heute prekäre Welternährungssituation entschärft werden kann. Die Gentechnologie könnte ihren Teil dazu beitragen, aber nur wenn sie vorsichtig genutzt wird. Die klassische Pflanzenzüchtung und die ökologische Grundlagenforschung dürfen jedoch auf keinen Fall vernachlässigt werden, weil auch sie das Potential zur Revolutionierung der heutigen Landwirtschaft in sich tragen.

ACKNOWLEDGEMENT

Fig. 10: Security guard watches by anti – GMO demonstration in 2008 (S. Zeller)

First of all, I would like to thank Bernhard Schmid for being a great doctor father: you keep on amazing me with your innovative ideas for new experiments, your statistical knowledge and your stamina to edit my manuscripts. Thank you for motivating me again after very setback – just think of my shock after the destruction of my experiments by activists in the first year or the hardship with the publishing of our first manuscript… You guided me well through the political mess of the GM controversy and taught me how easy communication is if you stick to nothing but the truth. You are a great mind and a wonderful person.

I would like to thank Olena Kalinina who worked with me on the same project. Your ability to work hard in the field and your statistical and ecological knowledge are remarkable. I wish you all the best in the future.

I would like to thank Beat Keller for getting the permissions to do our field experiments. Furthermore you co-authored my first manuscript even though the results and how we interpreted them must have been difficult for your. Thank you for putting science before politics.

Susanne Brunner for producing our most interesting model plants and for making the molecular make up of "her" plants understandable for ecologists.

A big thanks goes also to Agroscope Reckenholz-Tänikon ART for setting up the field experiments and supporting us in many aspects. Among the ART team I would like to thank the following people in particular:

… Carolin Luginbühl, for being the good soul of the field experiments.
… Andrea Foetzki, for organizing the field trial so nicely even though we arrived with very unconventional ideas from time to time.
… Michi Winzeler and Franz Bigler for explaining us the basics in agronomy.
… Paul Steffen, Director of the ART for many encouraging handshakes in the field.
… several anonymous Securitas guards who protected our experiment day and night and showed great interest in our research. Especially the one who helped us several times with the clean up and once even performed some measurements.
… the Zurich city police for not arresting us on their regular patrols around our field site although we might sometimes have looked like anti-GMO activists ourselves.

I would like to thank all members of "Wheatcluster.ch" for the good spirit on the field site and their truly interdisciplinary work. Especially, Yi Song, Andreas Lindfeld, Gerri Herren, Ale Fammartino and Joana Meyer for your companionship.

A special thank you goes to Petra Maria Bättig-Frey, the communication person of the consortium. Thank you for giving me the opportunity to communicate my research to the public and for the endless discussion about the benefits and risks of GM plants. I will never forget your courage when we both travelled to Basel to join a podium discussion organised by of anti-GMO activists of the worst sort.

I would like to thank Fabio Mascher for showing us his plant diseases in Pully.

I would like to mention the Swiss National Science Foundation for the funding of the important National Research Program NRP 59 and our project in particular (SNF 405940-115607). In this aspect I would also like to thank the Swiss parliament that spoke the money for this research program soon after the public voted for a strict GM moratorium and of course the Swiss tax payers for financing us.

I would like to mention Greenpeace for their nerves to fight our field experiments persistently (non-violently at least) and nevertheless citing our papers once they were published.

Thanks to Silvan Rieben and Simone Nägeli for being great master students. You allowed us to complete several experiments that we had not time to do ourselves and gave me the opportunity to practise my supervising skills.

I would like to thank Simone von Burg for her academic and political discussions at the IfU and in the field.

Yurij Kostetski, the husband of Olena helped us tremendously with his muscle power and stamina that allowed us to seed our field experiments, gather incredible amounts of data. I would like to thank him especially for spending several months in the glasshouse and doing nothing but threshing, weighing and counting millions of seeds.

Thanks to Mireia Nuñez-Marce, our Erasmus student from Spain, to help me with the first experiments in the glasshouse and the field.

I would like all harvest helpers who assisted us in the field: Adele Ferrari, Alicia Argüello García, Ana Suárez, Andreas Kundela, Angela Pauletto, Angelica Lopez, Daniel Trujillo Villegas, Debra Zuppinger-Dingley, Dominique Keller, Ellen Annika Waibel, Florin Ammann, Jorrit Nico Bachmann, Juliana Nates Jimenez, Luisa Last, Marc Schmid, Martin Baruffol, Matthias Zeller, Peter Schmid, Pirmin Scheuber, Sara Bischof, Stephanie Hartmann, Silvia Mathis, Tobias Züst, Tugce Arslan, Valeria Moncada Martinez, Veronica Iniguez and possibly more. Thank you for your great flexibility. I sometimes had to call you up in the middle of the night to ask you to start working at 6:00 in the morning the next day. You worked hard even in the hottest summer days and I will never forget the beers and the swimming in the Katzensee.

I will never forget the intellectual support of the "Old Masters" Atlant Bieri, Andres Overturf and Dan Tamir. You helped me with press releases and blog articles that were never published and supported my simultaneous fight against anti-GM activists and the pro-GM lobby. A special thank goes to Dan for being a great friend and for providing me with urgently needed inspiration in the final weeks of my thesis writing.

I would also like to thank the "Bios", my true friends, for listening to my stories and supporting my fight for the truth. Thanks Alex for finding spelling mistakes such as "GM – cops" and "pant diseases" at our Bios – weekend in Appenzell.

Thanks to Susann Eichenberger for the reviewing the German parts of this thesis.

The IEU provided a wonderful work atmosphere. I would like to thank Isabel Schöchli, Lilli Strasser, Maja Weilemann, Theres Zwimpfer, Georg Feichtinger, Pascal Niklaus, Helmi Brandl, Jana Petermann, Xuefei Li and everyone else for your help.

Finally, all of this would not have been possible without the support of my parents, Markus and Isabella Zeller and Annatina Schmidheiny. There are no words great enough to thank you!

CURRICULUM VITAE

Name:	ZELLER
First Names:	Simon Lukas
Date of Birth:	06.11.1982
Citizenship:	Gossau SG
Place of Birth:	Frauenfeld, Switzerland

Work Experience

06.2007 – 05.2011	PhD Thesis, Zurich Life Science Graduate Program Ecology "Ecology of genetically modified wheat: performance, resistance costs, mixture effects and gene flow " University of Zurich Institute of Evolutionary Biology and Environmental Studies Prof. Bernhard Schmid
07.2005 – 09.2005	Intership at the EMPA St. Gallen

Education

10.2005 – 03.2007	Specialized Master in Environmental Sciences "Host-plant selectivity of rhizobacteria in a crop/weed model system" University of Zurich Institute of Environmental Sciences: Environmental Microbiology Dr. Helmut Brandl and Prof. Bernhard Schmid
10.2002 – 07.2005	Bachelor in Biology, Geography (2. Subject) University of Zurich
01.2001 – 07.2002	Kantonsschule Romanshorn, Maturity Exam
01.2000 – 12.2000	High School Exchange Year in Manitoba, Canada

Publications

Zeller S., Brandl H., Schmid B. (2007) Host-plant selectivity of Rhizobacteria in a crop/weed model system. *PloS ONE* **2/9**, e846 doi:10.1371.

Zeller S., Kalinina O., Brunner S., Keller B. & Schmid B. (2009) Transgene x Environment Interactions in Genetically Modified Wheat. *PLoS ONE* **5/7**, e11405. doi:10.1371.

Brunner, S., Hurni, S., Herren, G., Kalinina, O., von Burg, S., Zeller, SL., Schmid, B., Winzeler, M. & Keller, B. (2011) Transgenic Pm3b wheat lines show resistance to powdery mildew in the field. *Plant Biotechnology Journal,* **9**, 897–910.

Reviewing

Reviewing

Faculty of 1000, Assistant Faculty Member (since 2010)
Reviewer for the following scientific journals:

- Journal of Applied Ecology
- *European Journal of Plant Pathology*,
- Perspectives in Plant Ecology Evolution and Systematics
- African Journal of Biotechnology
- Flora

Book Reviews: Beat Glogger, 2009; Peter Brandt, 2009

VSG
VDM Verlagsservicegesellschaft mbH

Die VDM Verlagsservicegesellschaft sucht für wissenschaftliche Verlage abgeschlossene und herausragende

Dissertationen, Habilitationen, Diplomarbeiten, Master Theses, Magisterarbeiten usw.

für die kostenlose Publikation als Fachbuch.

Sie verfügen über eine Arbeit, die hohen inhaltlichen und formalen Ansprüchen genügt, und haben Interesse an einer honorarvergüteten Publikation?

Dann senden Sie bitte erste Informationen über sich und Ihre Arbeit per Email an *info@vdm-vsg.de*.

Sie erhalten kurzfristig unser Feedback!

VDM Verlagsservicegesellschaft mbH
Dudweiler Landstr. 99
D - 66123 Saarbrücken

Telefon +49 681 3720 174
Fax +49 681 3720 1749

www.vdm-vsg.de

Die VDM Verlagsservicegesellschaft mbH vertritt

VDM Verlag Dr. Müller
LAP LAMBERT Academic Publishing
SVH Südwestdeutscher Verlag für Hochschulschriften

MIX
Papier aus verantwortungsvollen Quellen
Paper from responsible sources
FSC® C105338

Printed by Books on Demand GmbH, Norderstedt / Germany